# Preservation of Meat and Poultry Products

# Preservation of Meat and Poultry Products

*– Author –*
*Vishnu Yadav*

Wordpen Academics
Publishing Science That Matter

**ISBN: 978-93-90438-51-8 (Hardback)**

**Published by :** **WORDPEN ACADEMICS**
**4736/23, Ansari Road, Darya Ganj,**
**New Delhi – 110 002**
**Phone: +91-011-35004336**
**E-mail: info@wordpenacademics.com**

***Printed at*** : **Replika Press Pvt. Ltd.**

*Exclusive Distributor:* ***Bio-Green Books, New Delhi***

# Preface

The history of food preservation employed by humans is fairly old. Thousands of years ago, people had already used different techniques without actually understanding the scientific principal behind them. Some historians describe two major periods in terms of food consumption. The first is called the food-gathering period that covers the time from human origin, over one million years ago, to 8,000–10,000 years ago. The second is called the food producing period that continues until today. It appears that food spoilage problems were encountered early in the second period when people started to produce and store their own food for extended periods of time. Spoilage problems caused by improper storage as well as disease transmission via food have required some innovations/solutions. Drying food was one of the earliest methods employed to store foods like grain and fruits and was also used to preserve slices of meat. Sun-dried grain and meat could be stored for extended periods of time. Some cultures discovered that drying and smoking meat over fire could substantially extend the shelf life of the product. Later, fermentation of grains resulted in the production of beer, which can be traced back to ancient Babylonia at around 7000 BC. Other people, such as the Samaritans, are believed to have been the first great livestock breeders and dairymen, who were among the first to make butter at about 3000 BC. They were also known to use salted meat, fish and dried skins. The early Egyptians in 3000 BC were known for their ability to ferment dairy products and make cheeses. Salted meat was known to be used by the Israelites, the Chinese and the Greeks; the latter also passed it to the Romans. Evidence of sausage fermentation by the ancient Babylonians and the ancient Chinese people goes as far back as 1500 BC. It is doubtful whether people at that time understood the nature of food preservation by fermenting microorganisms; however, they used it fairly successfully. It was probably common to "seed" a new batch with material from a successfully previous batch and by that transmit the "right culture" to reduce chances of spoilage.

Raw meat and poultry products should be maintained at 40° F or below to greatly reduce the growth rate of any pathogenic bacteria that may be present on their surfaces. Chilling is required of all raw product unless it moves directly from the slaughter line to heat processing or cooking, which destroys pathogens. Meat and poultry products are chilled immediately after slaughter to acceptable internal temperatures which insure the prompt removal of the animal heat and preserve the wholesomeness of the products. Generally, red meat carcasses (which are above 90° F at the time of slaughter) are chilled in a blast cooler with rapidly moving chilled air, and, in some instances, a cold water shower. Poultry is required to be chilled to 40° F or less within specified time frames, depending upon the size of the carcass. Whole birds and parts of major size are chilled in ice or ice and water media. Poultry parts are chilled in ice, air or water spray with continuous drainage. Giblets must be chilled to 40° or below within two hours of slaughtering the birds. Packaging is a physical barrier to cross contamination. Microorganisms exist everywhere in nature. They are in the soil, air, and water. The simple act of covering food keeps microorganisms from contacting the food. Covered perishable foods can be stored longer and at better quality than uncovered foods. Modified atmosphere packaging (MAP) and vacuum packaging help prolong storage (see following question). Refer to the charts on pages 5 and 6 for recommended storage times.

The book provides a clear, scientific and technical understanding of this subject. The information contained herein will equip the readers with latest and up-to-date knowledge.

*—Vishnu Yadav*

# Contents


| | | |
|---|---|---|
| | *Preface* | *v* |
| 1. | Introduction | 1 |
| 2. | Meat Preservation by Thermal Treatment | 36 |
| 3. | Basic Methods of Quality Control | 51 |
| 4. | Egg Packaging, Transport and Storage | 152 |
| 5. | Safely Preserving Eggs | 184 |
| 6. | General Hygiene Principles for Meat Handling | 204 |
| 7. | Dry Sausages | 282 |
| | *Bibliography* | 317 |
| | *Index* | 319 |

# 1

# Introduction

Although the demand in developing countries for animal proteins is increasing, animal production has failed to keep pace with the growth in demand and to make full use of its potential in developing countries.

The nutrition of resource-poor rural producers can be improved both directly by consumption of animal food products or indirectly by enabling the purchase of food with returns from animal product sales.

The average per caput supply of meat in developing countries is very low. In 1979 and 1986 it amounted to 12.7 and 15.4 kg/caput/year respectively as compared to the developed countries with 74.3 and 77.9 kg/caput/ year respectively. The average per caput supply in developing countries in Africa reached the average level of all developing countries in 1979 with 12.9 kg/caput/ year but in 1986 remained at 11.8 kg/caput/ year, below the level of the developing countries combined and even below the 1979 level.

While efforts are increasing to support animal production in developing countries, they are not matched by similar efforts to use preservation to overcome seasonal variation in meat supply. In addition the existing conditions for slaughtering and meat handling in rural areas which cause quality deterioration and post-harvest losses of meat- and food-borne diseases in consumers must be improved.

In fact there is a lack of effort to provide knowledge and skills in adequate hygienic slaughtering, meat cutting and handling under rural conditions. Taking into account that an uninterrupted cold chain for meat cannot be expected in many developing countries in the near future, the absence of meat preservation techniques presents a serious constraint to the development of viable meat production by resource-poor rural livestock producers.

Adequate meat preservation complements a marketing system which by necessity has been adapted to a fast throughput of fresh meat and which does not facilitate the use of surplus meat in periods of meat shortage.

## Slaughtering and Raw Materials for Meat Preservation

Product quality and shelf-life of preserved meat and meat products depend on the microbiological and biochemical status of the carcass meat used for processing. In other words, the raw material must be as clean as possible and derived from slaughter animals in good and healthy condition.

Obviously the way of slaughtering animals plays an important role. Many irreversible quality losses, especially with regard to the hygienic quality, originate from improper slaughtering and carcass handling. Slaughter techniques, particularly under rural conditions, cannot and need not be sophisticated, but they must allow carcass meat which complies with basic hygienic requirements to be obtained whether slaughtering at the farm, or in slaughterslabs or slaughterhouses.

Careless slaughter and meat handling result in:

- improper and insufficient bleeding of the animals, leaving a relatively high amount of blood in the muscle. As a consequence the meat does not reach the necessary degree of acidity and its shelf-life is reduced;
- improper dehiding of the carcass, leading to heavy contamination of the meat surfaces by frequent contact with the personnel, polluted floors and dirty tools and equipment;
- improper evisceration through accidental opening of stomachs and tripes, leading to contamination spreading to internal and external surfaces of the carcass;
- contamination of the meat when carcasses are split on the ground;
- contamination of the meat during transport, when carcasses or parts of them make contact with unclean clothes, hooks, bars or containers;
- contamination of the meat stored under unhygienic conditions and because of faulty handling in meat markets, shops and processing plants.

It has to be borne in mind that faulty meat handling, apart from affecting the quality and shelf-life of meat and processed products, may also endanger the health of consumers. Massive contamination will not only enhance meat deterioration caused by food-spoiling bacteria, but may also cause poisoning of consumers by toxin-producing micro-organisms. Toxin-producing micro-organisms find a favourable environment not only in fresh meat stored for a prolonged period at ambient temperatures, but also under certain circumstances in processed meat or even in preserved meat. Appropriate hygienic measures during slaughtering and meat handling are therefore indispensable.

A proper way of improving the slaughter hygiene is to carry out as much of the slaughter operations as possible with the carcass in a hanging position.

## Rural Slaughtering on the Farm

Slaughter operations of small animals can easily be performed with the carcass in a hanging position. Problems may arise when slaughtering large animals such as cattle. The bleeding of the animal and cutting off the hind feet is done on the ground and the opening of the skin of the belly and leg region may also have to be done in this position. To complete skinning, evisceration and splitting, simple wooden structures and ropes to gradually hoist up the carcass are strongly recommended.

Remarkable improvements in slaughter hygiene can be achieved by this technique with much less visible contamination (dirt) and a lower degree of invisible contamination. The structure built for slaughtering can also be used for cutting and deboning the carcass.

### *Slaughterslabs*

Slaughterslabs are used in rural communities where the throughput of animals is very small, but where slaughtering in the field should be avoided. Animals should be suspended from a wooden or metal frame using a gantry hoist to ensure that carcasses are kept off the floor. The simplest slaughterslab consists of a concrete platform with gantryhoist facilities.

The platform should be approximately 50 cm above the ground and sloped to a drain. A roof should be added as protection against adverse weather conditions. Slaughterslabs should be fenced and roofed and have facilities for hygienic slaughtering and adequate disposal of effluents. If pigs are to be killed a scalding vat should be provided.

## Fao's Small-scale Modular Slaughterhouse

The Animal Production and Health Division of the Food and Agriculture Organization of the United Nations (FAO) has developed a model project for village meat industry in which one of the main components is a smallscale modular slaughterhouse. The design incorporates the use of locally available construction materials and unsophisticated equipment. The slaughterhouse can be built in modules, adding units to the central slaughterhall for operations such as by-product utilization, meat preservation, processing and butchering. Designs have also been prepared for the construction of a meat market in order to facilitate the integration of production, processing and marketing.

***Figure:*** *Simple wooden structure to hoist slaughter animal after bleeding.*

It is intended that from the basic nucleus a number of small-scale industries be developed consecutively, e.g. meat preservation by low-cost technologies, traditional tanning, handicrafts based on hides, skin, bone and horns, etc. The objective is to increase employment, in particular for rural women, create market outlets for livestock products and increase the income of small producers.

The project could become a focus for farmer organization and a centre of technical assistance for producers. The organization of small producers is indispensable to ensure their participation and for the success and continuity of the project.

FAO is promoting the establishment of slaughterhouses of this type in some rural areas because this would increase the availability and quality of meat, improve the utilization of by-products, avoid rural migration and increase employment. The slaughterhouses may, therefore, constitute a tool for rural development, increasing the income level in the countryside and villages, thus improving the living conditions of the rural communities.

Rural slaughterhouses can also facilitate veterinary control of livestock (ante-mortem inspection) and carcass meat (post-mortem inspection). Slaughtering can be organized to follow a determined daily or weekly time schedule for which compulsory veterinary controls can be arranged. In addition, adequate technical facilities for efficient meat inspection can be provided, i.e. vertical position of the carcasses and special platforms for the personnel. Special hooks or inspection tables can be provided for the inspection of internal organs.

The detailed design for a small-scale modular slaughterhouse was developed by FAO. It includes designs, specifications, and schedule of quantities for a slaughterhouse and meat market suitable for small communities.

Provision is made for slaughter of all species: cattle (or buffalo), sheep, goats and pigs, though because of space limitations, concurrent slaughter of different species is not possible. The abattoir capacity will be dependent on the mix of animals being slaughtered. Daily throughputs of approximately 10 large stock (e.g. cattle) or 50 small stock (sheep, goats or pigs) or a combination thereof could be achieved with this design.

The facilities are divided into modules which can be combined as required to suit a particular location.

The following modules are included:

- Lairage
- Slaughter floor
- Chiller
- Tripe room
- Meat cutting and processing room
- Solid waste and blood disposal
- Hides and skin processing
- Effluent disposal
- Electric light and power
- Water supply
- Meat market.

A possible overall abattoir layout based on these modules is shown in figure. This layout shows a typical arrangement for a facility designed to handle beef, small ruminants and pigs. Modifications for a large beef kill and/or the elimination of pig slaughter (e.g. for Muslim communities) are possible.

It is evident that when establishing slaughterhouses each country or even separate localities must adopt a solution incorporating special local conditions, locally available materials and manpower, etc.

The designs foresee procedures for slaughter of each species as follows:

### Cattle

The animal is led into the bleeding area where it is restrained by a tether through the floor ring prior to stunning (using a captive bolt pistol). After stunning the animal is shackled by one leg and hoisted with a rope pulley block. The animal is then stuck and allowed to bleed in this position and the blood collected in a drum for disposal.

Once bleeding is complete the head can be removed and the animal lowered on to the cradle for dressing. It is also possible to dress the hanging animal.

The feet are then removed, the skin opened up along the breastbone and the hide partially flayed. Leg hooks are attached and the carcass raised to a half-hoist position on the spreader. Flaying can then be completed and the hide removed. The paunch can then be removed to the inspection buggy and the red offal (including lungs if treated as edible) placed on hooks or the inspection table for inspection.

After inspection the carcass can be split and quartered, the quarters being individually hung on the low rail.

Once the carcass has been partially flayed and half-hoisted a second animal can enter the bleeding area.

### Pigs

Pigs are first stunned in the stunning area then hoisted for sticking and bleeding and then transferred to the scald tank. After scalding for approximately five minutes at 60°C the carcass is removed to the scraping table.

After scraping a gambrel can be inserted into the hind legs and the carcass transferred to the overhead rail for final scraping and evisceration. Once a pig is clear of the scraping table the next pig can be placed in the scalding tub.

### *Sheep and Goats*

These would be slaughtered and dressed on the rail in the pig area in a similar manner to pigs. The scraping table is removed to one side during processing of sheep and goats.

## Carcass Handling

After slaughter carcasses should be chilled in chilling rooms. Chilled meat is a requirement for many methods of further processing. However, in rural areas of developing countries refrigeration facilities are generally lacking. Before cutting, carcasses should therefore be carefully examined for signs of taint caused by microbial spoilage.

Unchilled meat as a raw material is suitable for dried meat and certain meat products which undergo a heat treatment immediately after processing.

When dealing with hot carcasses, cuts and trimmings should be either consumed or processed (dried) on the day of slaughtering. It is obvious that under these conditions good hygiene during slaughtering and meat handling is of great importance for the quality of the final product. The higher the initial contamination, the faster the meat deterioration, especially under high ambiental temperatures.

In this context a new way of short-term meat preservation that could be especially beneficial for prolonged periods of handling of carcasses or meat cuts during transport, etc. should be mentioned. Meat surfaces are treated with organic acids such as acetic, lactic, citric, tartaric and ascorbic acid, as well as sodium sorbate.

These compounds are from different natural foods and not toxic and may be used alone or in combination as dipping solutions or sprays on the surface of meat and meat products. Treatment of carcasses with these products has proved to be successful under conditions in developing countries. With an aqueous solution of 20 percent sodium sorbate, 5 percent sodium acetate and 5 percent sodium chloride sprayed on warm beef carcasses, the shelf-life of meat at 25° to 35°C was doubled. However, more work is needed on the subject, especially for tropical conditions.

### *Cutting*

As whole carcasses of beef or pork are too large to be easily transported in one piece, they are split into sides or cut into fore- and hindquarters. The development of meat processing introduced the need for cutting quarters, halves or whole carcasses into smaller pieces which, according to their quality and market value, are used for culinary purposes and processing respectively.

The culinary meat in the form of primary cut is mostly sold to wholesale dealers who bone and cut it into sub-primals and finally into retail cut. The point at which wholesale cutting ends and retail cutting begins is not clearly defined. In the traditional, small, independent butchers' shops these operations may take place on a cutting table behind the counter in order to give the butcher the maximum opportunity to select the most suitable piece of meat and prepare it in a manner which is most likely to satisfy the consumer.

Meat for processing comprises parts of lower quality but also high-quality meat for the manufacture of special products like hams, smoked pork loins and dried meat. It can be divided into different classes according to the amount of fat and connective tissues.

There are several regional, country, and local differences in cutting the animal carcass into primary or retail cut and butchery practices vary depending on geographical location, tradition and habits, and demand for high-quality meat.

For the preparation of dried meat, two methods have been recommended. In both, carcasses are divided into two sides along the backbone. Each side is cut crosswise into two quarters which are divided into primal and retail cuts, and then boned and trimmed.

The pistola cutting system, with the side of beef quartered between the fifth and sixth ribs, allows the complete separation of all first-quality meat cuts in the hindquarters and loin regions. After separation of the forequarters, the flank piece is removed by freeing the muscles of the abdomen from those of the proximal pelvic limb. Separation is completed by extending the cut down the side, parallel to and at a distance of some 20 cm from the backbone.

The hindquarter piece, which contains all the first-quality muscles, may be separated into the hind leg, rump and loin cuts by cutting and sawing directly across the *Bicepsfemoris* at a point just below the exposed point of the pelvic bone. The forequarter is similarly divided into two by removing the foreleg and blade cut from the thorax and neck. According to the second method the carcass is split into sides, along the spinal column. After that, the sides are cut horizontally into two quarters: hindquarters and forequarters. The cut line runs immediately after the last rib, which excludes leaving any ribs on the hindquarters. The hindquarters are hooked by the Achilles tendon and the forequarters by the last two ribs.

In both cutting systems further division into commercial or retail cuts is usually necessary in order to separate the muscles or groups of muscles most suitable for drying from those which, by nature of their characteristics and retail value, should be sold fresh or processed by other methods.

***Table:*** *Pistola system Primal cuts expressed in percentage of cold carcass weight*

| ***Primal cut*** | ***Percentage*** |
|---|---|
| Flank and rib | 14 |
| Hind leg | 29 |
| Rump and loin | 20 |
| Foreleg and blade | 15 |
| Thorax and neck | 22 |

***Source***: ILO(1985)

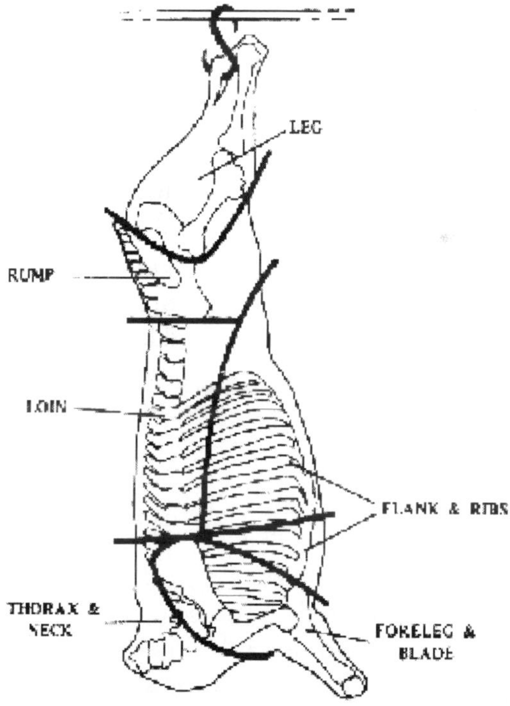


***Figure:*** *Beef carcass showing primal cuts according to pistola procedure.*

After boning, the meat is trimmed to improve the quality of the finished product. Trimming consists of removing all or part of the extra fat layers, coarse tendons, fascia, nerves and blood vessels and, where necessary, bits of muscle which have been improperly cut or superficially damaged. Infiltrated fat, particularly melted fat, must also be removed to ensure a high-quality finished product which keeps well. Not only does this fat have a peculiar taste, it is sticky, unpleasant, easily becomes rancid and often forms a barrier which can interfere with the drying process.

The bones can be dried as well. They should be cut into smaller pieces on a clean clog. However, care should be taken to leave the connective tissue and ligaments intact, so that the pieces of the bones can be suspended for drying after having been dipped into a salt solution.

## Some Important Non-meat Ingredients For Meat Processing

*Salt* is an important ingredient in the preparation of meat emulsions for imparting the typical flavour of processed meats and to contribute to keeping quality. The salt content of most processed meats ranges between 2.5 to 5.0 percent of the final product. A higher salt content would produce a salty taste. To function as a preservative, salt concentrations in the product of about 17 percent would be necessary, far too high for a palatable product.

*Nitrates and nitrites* are curing ingredients. The development of the red or pink colour is their most obvious effect. In addition, nitrites:

- affect the flavour of the products through their action as anti-oxidants; and
- have bacteriostatic properties, particularly in canned products (sodium nitrite is an effective inhibitor of the growth of the bacterium *Clostridium botulinum*).

Nitrate in itself is not effective in producing the curing reaction until it is first broken down into nitrite. This is a slow process which depends upon the presence of bacteria in brine and meat. Therefore nitrate has largely been replaced by nitrite.

Nitrite provides the ultimate source of the nitric oxide that reacts with the myoglobin pigment of the muscle tissue. Levels in excess of 200 parts per million of sodium nitrite should not be used. Since nitrate and nitrite are added in small amounts, it is recommended that they are first dissolved in water to ensure uniform distribution.

*Extenders* in general are added to lower-quality products for economic reasons. Some of them improve binding properties, cooking yields, slicing characteristics and flavour. The ones most often used are:

- dried skim milk and milk proteins;
- various cereal flours such as wheat, rice, oats, corn;
- soy products such as flours (containing about 50 percent protein), grits (which are similar to soy flour in composition but they are larger in particle size and more adaptable to meat products), textured soy protein (similar to grits except that the texture is changed to more closely resemble the texture of ground meat), soy protein concentrates (containing about 70 percent of protein, that is available either in a coarse granular form or as a flour), soy protein isolates (containing about 90 percent protein), useful both as binders and emulsifiers.

Various non-meat proteins are being developed for use in sausage and processed meats. These include protein from oil seeds other than soya, such as cotton seed and peanuts, as well as from single-cell plant source such as torula yeast. Most extenders are usually limited to 3 percent in the dry state or to 10 percent after swelling.

*Seasoning* is a comprehensive term for ingredients which improve the flavour of processed meats. Salt and pepper form the foundation upon which many seasoning formulae are built. Other ingredients such as spices, herbs and vegetables are supplementary, although necessary to obtain the distinctive flavour associated with various products.

Spices are aromatic substances of vegetable origin and include cinnamon, cassia, clove, ginger, mace, nutmeg, paprika, pepper, cardamom, coriander and mustard. They vary in composition. The aromatic and pungent components which render them valuable are present in volatile oils and resins. Some success has been achieved in extracting these components. Spice extracts offer the advantage of being easier to dose and to store. Condiment herbs include sage, savory, bay leaves, thyme and marjoram. The dried leaves of any of them can be used in the preparation of sausages and other meat products. Condiment vegetables are onion and garlic. All spices and seasonings should be stored under dry conditions and possibly in sealed containers. They should not be exposed to direct sunlight and should only be ground or crushed on the day of manufacture.

## Simple Techniques for Production of Dried Meat

### *Principles of Meat Drying*

Drying meat under natural temperatures, humidity and circulation of the air, including direct influence of sun rays, is the oldest method of meat preservation. It consists of a gradual dehydration of pieces of meat cut to a specific uniform shape that permits the equal and simultaneous drying of whole batches of meat.

Warm, dry air of a low humidity of about 30 percent and relatively small temperature differences between day and night are optimal conditions for meat drying. However, meat drying can also be carried out with good results under less favourable circumstances when basic hygienic and technological rules are observed. Intensity and duration of the drying process depend on air temperature, humidity and air circulation. Drying will be faster under high temperatures, low humidity and intensive air circulation.

Reducing the moisture content of the meat is achieved by evaporation of water from the peripheral zone of the meat to the surrounding air and the continuous migration of water from the deeper meat layers to the peripheral zone.

There is a relatively high evaporation of water out of the meat during the first day of drying, after which it decreases continuously. After drying the meat for three or four days, weight losses of up to 60–70 percent can be observed, equivalent to the amount of water evaporated. Consequently, moisture losses can be monitored by controlling the weight of a batch during drying.

Continuous evaporation and weight losses during drying cause changes in the shape of the meat through shrinkage of the muscle and connective tissue. The meat pieces become smaller, thinner and to some degree wrinkled. The consistency also changes from soft to firm to hard.

In addition to these physical changes, there are also certain specific biochemical reactions with a strong impact on the organoleptic characteristics of the product. Meat used for drying in developing countries is usually derived from unchilled carcasses, and rapid ripening processes occur during the first stage of drying as the meat temperature continues to remain relatively high. For that reason the specific flavour of dried meat is completely different from the characteristic flavour of fresh meat. Slight oxidation of the meat fats contributes to the typical flavour of dried meat.

Undersirable alterations may occur in dried meat when there is a high percentage of fatty tissue in the raw meat. The rather high temperatures during meat drying and storage cause intensive oxidation (rancidity) of the fat and an unpleasant rancid flavour which strongly influences the palatability of the product.

Meat drying is a complex process with many important steps, starting from the slaughtering of the animal, carcass trimming, selection of the raw material, proper cutting and pre-treatment of the pieces to be dried and proper arrangement of drying facilities. In addition, the influence of unfavourable weather conditions must also be considered to avoid quality problems or production losses. The secret of correct meat drying lies in maintaining a balance between water evaporation on the meat surface and migration of water from the deeper layers.

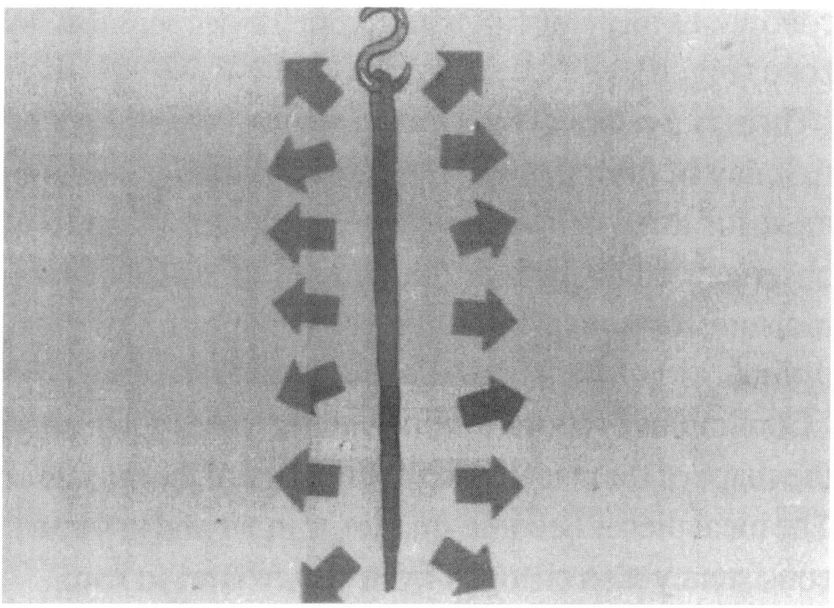

***Figure:*** *Dehydration process of a piece of meat suspended under drying conditions (schematic).*

In other words, care must be taken that meat surfaces do not become too dry while there is still a high moisture content inside the meat pieces. Dry surfaces inhibit the further evaporation of moisture, which may result in products not uniformly dried and in microbiological spoilage starting from the areas where the moisture content remains too high.

Adherence to the following meat-drying techniques should avoid failures in the production of dried meat and ensure obtaining products of good quality with a long shelf-life. The following description of the basic technology of meat drying includes the salting of the meat before drying. Presalting is not absolutely necessary, but has certain advantages, particularly for the drying of meat strips and large flat meat pieces and is therefore strongly recommended for this type of product.

### *Selection of Meat For Drying*

As a general rule only lean meat is suitable for drying. Visible fatty tissues adhering to muscle tissue have a detrimental effect on the quality of the final product. Under processing and storage conditions for dry meat, rancidity quickly develops, resulting in flavour deterioration. Dry meat is generally manufactured from bovine meat although meat from cameloids, sheep, goats and venison (e.g. antilopes, deer) is also used.

The meat best suited for drying is the meat of a medium-aged animal, in good condition, but not fat. Meat from animals in less good nutritional condition can also be used for drying, but the higher amount of connective tissue is likely to increase toughness. It is very important that raw material for the manufacture of dry meat is examined carefully for undersirable alterations such as discolouration, haemorrhagic spots, off-odours, manifestation of parasites, etc. Such defects must be trimmed off.

Carcasses have to be properly cut to obtain meat suitable for drying. Owing to their size, beef carcasses are more difficult to handle under rural conditions than carcasses of sheep, goats or game. In the absence of chilling facilities, beef carcasses must be cut and deboned immediately after slaughter.

## Beef Carcass Cutting, Trimming and Deboning

### *Carcass Cutting*

The carcass is first split into two sides along the spinal column and then cut into quarters. Fore- and hindquarters are separated after the last rib, thus leaving no ribs in the hindquarter. For suspension the hindquarter is hooked by the Achilles tendon and the forequarter by the last two ribs.

### *Trimming*

After the quarters are suspended so that they do not touch the floor or anything around them, they are trimmed. Careful trimming is very important for the quality and shelf-life of the final product. The first step is to remove with a knife all visible contamination and dirty spots. Washing these areas will spread bacterial contamination to other parts of the meat surface without cleaning the meat.

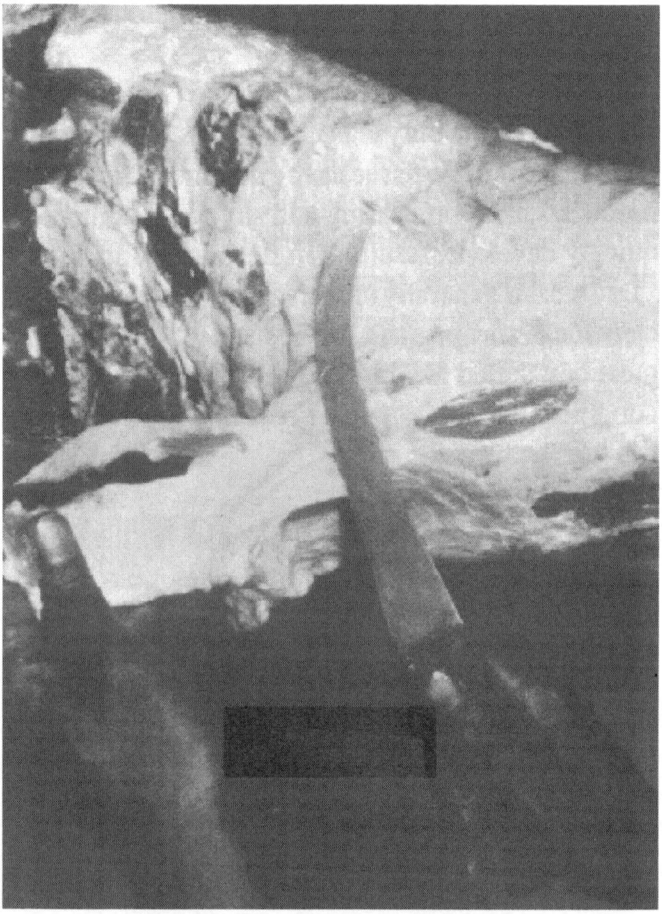

***Figure:*** *Trimming a beefcarcass.*

After completing the necessary cleaning of the meat surfaces, knives and hands of personnel must be washed thoroughly. Using a sharpened knife, the covering fat from the external and internal sides of the carcass and the visible connective tissue, such as the big tendons and superficial fasciae, are carefully trimmed off.

### Deboning

It is recommended that this operation should start with the hindquarters and follow with the forequarters. The aim is to remove the bones with the least possible damage to the muscles. Incisions into the muscles are inevitable but only at spots where the bones adhere and have to be cut off. Deboning of the suspended hindquarter should start from the leg and proceed to the rump and muscles along the vertebral column.

Deboning of the forequarter must start with cutting and deboning the shoulder separately, followed by cutting off the rib set, together with the intercostal muscles.

Deboning of the forequarter is completed by removing the meat from the neck and the breast region of the spinal column.

## Technique of Cutting Meat Pieces For Drying

Anatomic cuts, which were separated from the carcass, are suspended again and the big individual muscles are carefully cut out, while the smaller muscles are left together. The next step consists in cutting the muscles into thin strips. This operation is crucial for the appearance and quality of the final product. All strips to be dried in one batch must be cut to an identical shape. Care must also be taken to obtain rather long strips of meat.

There are two ways of cutting muscles or smaller muscle groups into strips:

- cutting the meat after placing it on an appropriate clean chopping board; or
- cutting the muscle in the hanging position.

In both cases the muscles have to be split exactly along the muscle fibres. The strips must be cut as uniformly and as smoothly as possible and the diametre of the strip must remain the same throughout the length. The length of the strips may differ, though it should not be less than 20 cm and not more than 70 cm. Meat cut into shorter strips requires considerably more time for hooking than the same quantity cut into longer strips. However, strips which are too long may break because of their weight.

Beef muscles suitable for drying are usually no longer than 50 cm (except the sirloin strip attached to the spinal column). However, strips longer than 50 cm can be produced by cutting the muscle along the fibre in one direction, without cutting through the end of the muscle. Using this technique long strips can be obtained, but their length should not exceed 70 cm for reasons of stability.

The thickness of the strips determines the duration of the drying process. Since thick strips take considerably more time to dry than thin ones, it is important that strips to be placed in the same batch are of the same cross-section, with only the length differing. Insufficiently dried or overdried pieces will be the result if this rule is not followed.

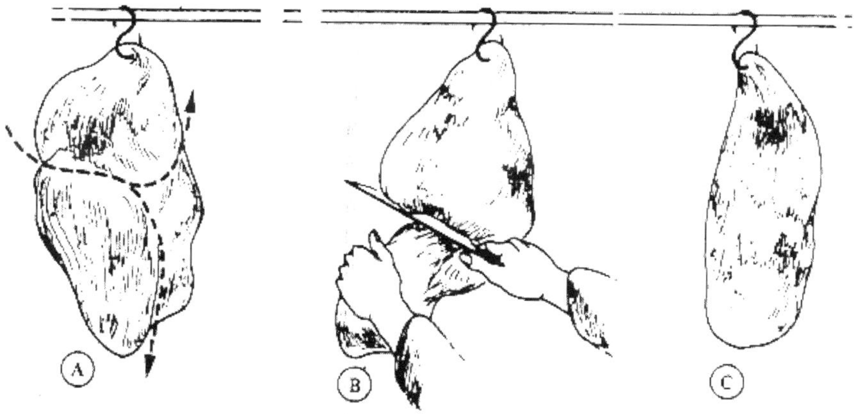


***Figure:*** *Suspended anatomic cut from the hindquarters ("silverside") (A) and splitting into individual muscles (B) which result in (C).*

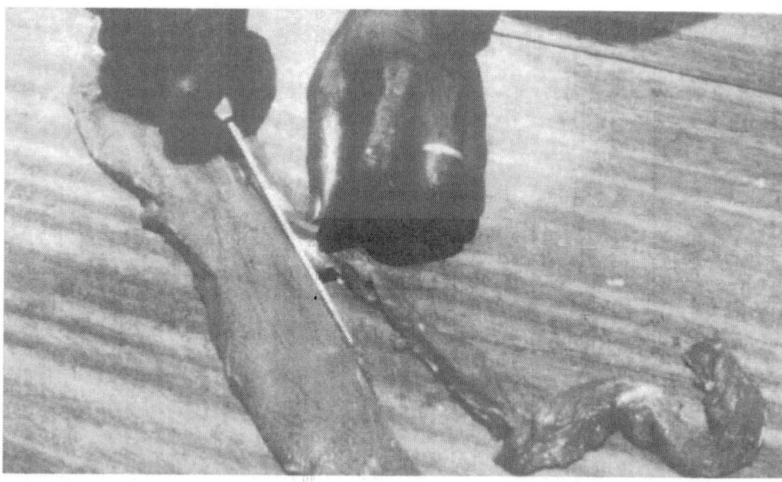

***Figure:*** *Cutting meat streeps from the muscle on a chopping board.*

Cutting muscles into long, thin and uniformly shaped strips requires experience and skill. Knives with broad blades are best suited for this purpose.

Under dry climatic conditions two basic shapes of meat pieces proved to be the most suitable for natural drying:

- strips with a rectangular cross-section of 1 x 1 cm; and
- flat-or leaf-shaped pieces with cross-sections of max. 0.5 cm x approx. 3, 4 or 5 cm.

## *Recommended Treatment Before Drying*

Because meat is always consumed slightly salted, the raw material may be presalted before drying. This procedure not only contributes to a more tasty product, but is also desirable from the technological and hygienic standpoint. Pure common salt is used for this purpose, either dry or dissolved in water. In the case of meat for drying cut into strips or flat pieces, the use of a 14-percent salt solution is preferred.

Dipping the meat into the salt solution serves first of all to inhibit microbiological growth on the meat surfaces. For that reason salting has to be carried out within five hours after slaughter, as after that period massive microbiological growth occurs which cannot be reduced by salt treatment.

Secondly, presalting is a protection against insects during drying. The freshly cut meat surfaces are very attractive to various insects, in particular domestic flies, which feed on the moisture excreted from muscle fibres. These insects cause considerable contamination of the meat and may also deposit their eggs into it. Meat is no longer such an attractive environment for insects after it has been dipped into the salt solution. The salt concentration on the meat surfaces keeps them away.

Furthermore, a thin layer of crystalline salt is formed on the surface of the meat during drying. The salt crystals are hygroscopic and absorb part of the water excreted from the meat, preserving the meat surfaces by keeping them dry. Dry meat surfaces inhibit the growth of bacteria and moulds which is one reason for the preservability of presalted and dried meat.

The salt solution is prepared by adding the necessary amount of edible common salt to water and dissolving it by intensive stirring. To obtain the recommended salt concentration of about 14 percent the amount of salt necessary for different volumes of water (expressed in litres) is indicated below:

| ***Water*** | ***Salt*** |
|---|---|
| (1) | (g) |
| 5 | 810 |
| 6 | 975 |
| 7 | 1 140 |
| 8 | 1 300 |
| 9 | 1 460 |
| 10 | 1 630 |

As soon as the salt is dissolved in the water, the meat strips are dipped into the solution, soaked for about five minutes and then drained. Draining should be done by placing the strips into a plastic sieve in order to allow the brine to drop off for collection and re-use.

The handling of the meat strips before drying has to be carried out under strictly clean conditions in order to avoid contamination and ensure a long shelf-life of the dried product. However, if accidental contamination of certain pieces occurs, further processing can only be undertaken with certain precautions. A special bucket with salt solution should be available in order to soak the contaminated pieces of meat, after having rinsed them previously in clean water. However, it must be borne in mind that the original quality of contaminated pieces cannot be restored. For that reason such pieces should always be dried separately, and not stored for a long period, but should be used as soon as possible in the preparation of meals.

### Methods of Suspending Meat Strips For Drying

***Figure:*** *Meat strips are soaked in a 14-percent salt solution for five minutes.*

The traditional way of suspending meat for drying by hanging strips over tree branches, wire or rope is not recommended because meat remains in contact with these supporting devices or may touch each other and thus not dry properly in these contact areas. Consequently, the chosen method should be to suspend the meat strips individually from one end, thus ensuring, through appropriate arrangement on the drying facility, free air circulation along the whole length of the pieces and fast and uniform drying. The contact of meat pieces with each other during drying must absolutely be avoided, since these areas will remain wet and humid for a prolonged period, thus making them a favourable environment for spoilage, bacteria and flies.

***Figure:*** *Draining the brine from the meat strips after soaking using a plastic sieve.*

The suspension of the meat strips can be done in different ways, either with hooks, loops or clips.

### *Suspension Using Metal Hooks*

This is a very simple but efficient way of suspending the meat strips. The meat strips are hooked at one end, always the thicker end for stability, and suspended on a horizontal wooden stick, tightrope or wire.

The metal hooks can easily be made, preferably from galvanized (non-corrosive) wire. Wire of 1 to 1.5 mm diametre is cut into pieces 15 cm long with a slanting cut so that the ends are sharp to allow piercing of the meat. In order to obtain an S-shaped hook, both ends of each piece are simply twisted around a circular stick.

### *Suspension Using Loops*

For this purpose a thin string or a somewhat stronger thread is best suited. The string is divided into pieces about 30 cm long with the ends knotted. The string is fixed to the stronger end of the meat strip by a double loop and pulled tight in order to prevent the meat from slipping out of the loop.

### *Suspension Using Metal Clips*

Clips 4 to 7 cm wide are best suited. They are easily placed on the stronger end of the meat strips. Whereas metal hooks and rope loops can only be used for the suspension of rectangular or similar shaped strips, the metal clips are very practical for the suspension of flat, leaf-shaped pieces. The special advantage of hanging leaf-shaped pieces by means of a clip is that the edges of the meat do not fold in during drying.

### *Installation For Drying Entire Batches of Meat*

It has already been pointed out that placing meat pieces for drying over wire, ropes or branches of trees is not recommended. Apart from problems of free air circulation under trees, some pieces may be intensively exposed to direct sun, whereas others are screened by the foliage. Furthermore, wind will transfer dust, twigs or leaves on to the meat and insects and birds will cause further damage.

A general disadvantage of this very simple method of meat drying is that it is practically impossible to shelter the meat in case of storm or rain.

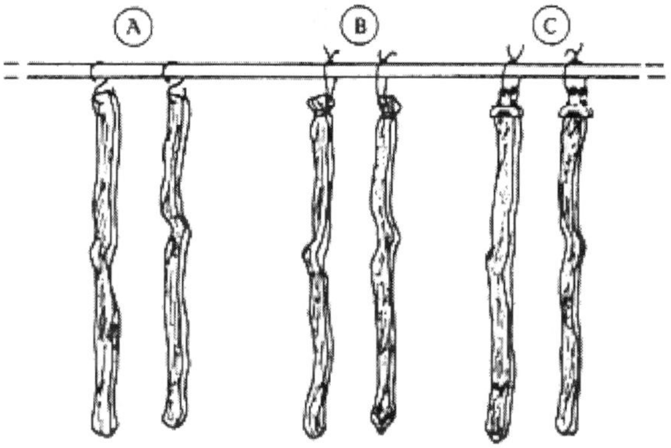


***Figure:*** *Suspension of meat strips on hooks (A), loops (B), and by means of clips (C).*

It is therefore recommended that natural meat drying be done by using simple premises and equipment, which can be made locally. The following descriptions give the different types of meat dryers.

These meat dryers are constructions of wood, metal and/or concrete, stationary or mobile, without or with a roof. For strips suspended by hooks or with a loop attached or fixed by clips, removable horizontal bars, either made of wood or metal or horizontal wire strings are needed.

## Sun Meat Dryer Made of Wood or Metal

This dryer consists of four wooden forks planted into the ground which are connected with two longitudinal, wooden traverses of about 4 m. Wooden or metal sticks for hanging the meat streeps are placed on the traverses at a distance of 15 cm from each other.

If there are not sticks available, strong metal wire or plastic rope can be spanned between the two longitudinal traverses and two additional transversal traverses should be fitted for reinforcement. Similar constructions, but with iron parts and traverses instead of the wooden ones are suited for the more industrial type of meat drying.

These constructions can also easily support a roof and meat drying can be done on two or more levels. However, care has to be taken that the traverses at the lowest level are not less than 1 m from the ground.

### *Mobile Meat Dryer*

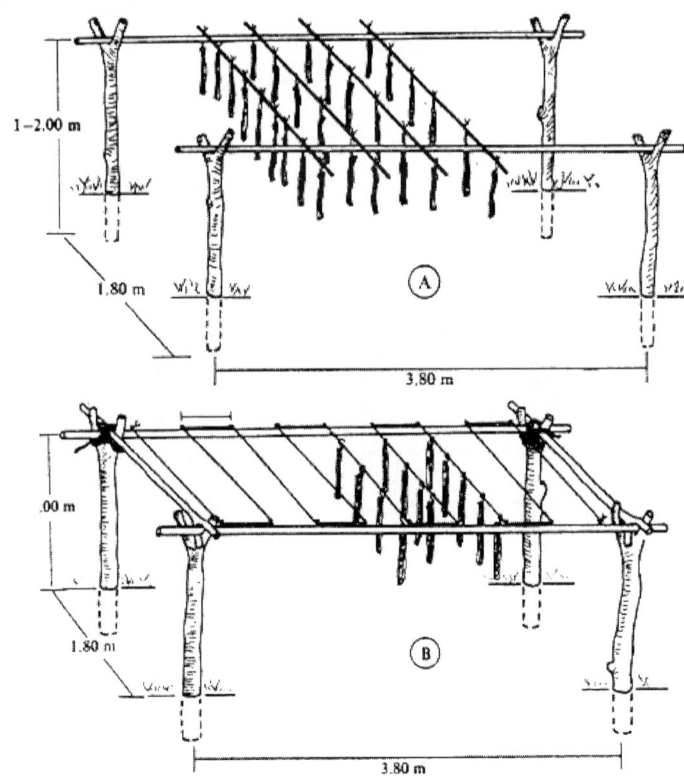


***Figure:*** *Simple wooden construction for meat drying using sticks (A) or wire/plastic rope (B) to suspend the meat strips.*

This type of dryer, which can be easily assembled or dismantled, can be moved to places where the animals are slaughtered. Although wooden constructions can be used for this purpose, for easy assembly and for a firm base metal constructions using 40-mm tubular iron bars are more convenient.

This type of dryer consists of two rectangular frames (2 x 2 m) placed 4 m apart from each other. They are connected with four longitudinal and eight transversal metal traverses in two levels from the ground (approx. 1 m and 2 m), thus permitting meat drying on the two levels. This dryer has the capacity for drying the meat of two beef carcasses at the same time. It is recommended that the upper level of the dryer be used for the suspension of meat from the hindquarters and the lower level for meat from the forequarters.

### *Meat Dryer with Protection Against External Influences*

In regions with strong and frequent winds, meat placed on the dryer must be protected from contamination by dirt, dust, sand, etc. In these cases it is recommended that the side walls of a roofed dryer (e.g. roof of corrugated aluminium) be covered with plastic foil up to a height of 0.80 m to 1.20 m from the floor. It is important to ensure that the upper parts of the dryer remain open for air circulation. Protection against insects is provided by covering the sides of the dryer with insect screen.

### *Arrangement of the Meat Strips in the Dryer*

In the dryer the meat strips are hung on horizontal plastic ropes, wires or sticks by means of hooks, loops or clips. As shown above, ropes and wires have often to be supported because of the heavy load of suspended meat, and they are not easily removable, which may cause difficulty in handling the meat to be dried.

Compared to flexible wires or ropes, firm wooden or metal sticks have proved to be the better solution for the following reasons:

- the sticks can be moved out of the dryer in order to suspend the meat pieces;
- the suspended meat pieces maintain the correct distance from each other on the firm stick and there is no need for supports for the sticks;
- sticks with the suspended meat can easily be moved, for example to a smoke house or for shelter in case of a storm or heavy rain; and

- weight losses can easily be monitored during drying by weighing the stick plus the meat without being forced to remove the meat strips.

The round wooden sticks can be made locally; they are usually 2 m long and 2 cm in diametre. This size permits the suspension of 25 to 30 pieces of meat. Metal sticks are in two different shapes, one tubular, made of galvanized water-pipes, and the other T-shaped. They are more expensive than wooden sticks but they last much longer.

The following rules for the arrangement of the meat pieces apply to both wooden or metal sticks:

- the number of meat pieces suspended must always remain the same for all sticks in the dryer (for instance 30 per stick), for reasons of optimal air circulation and easy control of theft, etc.;
- the distance between the individual suspended pieces must remain the same and be sufficient for free air circulation; and
- the longest and thickest pieces of meat have to be placed towards the ends of the stick whereas the thinnest and shortest pieces are kept in the middle, in order to expose the larger pieces to the stronger air circulation on the external part of the dryer.

### *Quality of the Finished Product*

Drying of meat of the shape described in this chapter takes four to five days. After this period the dried meat is ready for consumption and can be packaged, stored or transported. At this stage the product should meet the following quality criteria.

*The appearance* of the dried meat should be as uniform as possible. The absence of large wrinkles and notches indicates the desired steady and uniform dehydration of meat.

*The colour* of the surface, as well as of the cross-cut, should be uniform and dark red. A darker peripheral layer and bright red colour in the centre indicates incorrect, too fast drying, with the formation of hard rind which hinders evaporation from the deeper layers of the product.

In this case the central parts have a brighter colour and softer consistency and are, because of the higher water content, more susceptible to microbiological spoilage when packaged or otherwise stored. A softer consistency can also be recognized by pressing the meat with the fingers. These pieces should be kept for one more day in the dryer for finishing. The consistency of properly dried meat must be hard, similar to frozen meat.

*Taste and flavour* are very important criteria for the acceptance of dried meat by the consumer. Dried meat should possess a mild salty taste which is characteristic for naturally dried meat with no added spices. Off-odours must not occur. However, a slightly rancid flavour which occurs because of chemical changes during drying and storage is commonly found in dried meat. Dried meat with a high fat content should not be stored for a long period but used as soon as possible in order to avoid intensive rancidity.

Dried meat must be continuously examined for spoilage-related off-odour, which is the result of incorrect preparation and/or drying of the meat. Meat with signs of deterioration must be rigorously sorted out.

### Packaging and Storage

After taking the dried meat strips out of the dryer, a selection of the pieces based on length can be undertaken.

Packaging serves to protect the product from contamination to which the meat might be exposed on its way from the producer to the consumer.

Numerous materials are used for packaging dry meat, such as paper, plastic foils, aluminium foils, cellophane and textiles. The longest shelf-life is obtained using vacuum-packaging. Transparent plastic material and cellophane are more appealing to the consumer.

Packaging is employed for both the retail and wholesale trade. The weight per package of dry meat for retail sale usually does not exceed 1 kg, whereas those for the wholesale trade weigh 5, 10, 25, or 50 kg.

If plastic bags are used for packaging, the pieces of dry meat should be cut to a certain length so that they can be best arranged in the bags. Cardboard boxes are very useful for additional packaging.

During storage special care has to be taken to prevent dried meat, which is not packaged in water-proof containers, from becoming wet, resulting in rapid growth of bacteria and moulds. For this reason the premises for storing dry meat have to be rain-proof. It is further advisable to cover the piles of packaged dry meat with plastic sheets, as additional protection against moisture and dust. Dry meat protected in this way can be stored for more than six months.

During storage individual packages must be opened at least once a month and the organoleptic quality of the goods examined. These controls enable the persons responsible to evaluate storage conditions and to assess the shelf-life of the dry meat.

For controlling temperature and air humidity, it is useful to have a thermometre and hygrometre installed on the premises. A maximum-minimum thermometre is recommended to obtain the highest and lowest temperatures recorded between two readings. The temperature and relative air humidity should be carefully registered bearing in mind that dry meat is extremely sensitive to changes in environmental conditions, especially of the ambient temperature and relative humidity.

### *Preparation of Dried Meat For Consumption*

Dried meat manufactured as described above has to be rehydrated to resemble fresh meat again. Rehydrated dried meat has almost the same nutritive value as fresh meat.

Rehydration is in most cases combined with cooking. The procedure usually starts by putting the dried meat, which may be cut in smaller pieces, into a pot. The meat in the pot is then covered with water and boiled. The rehydrated and cooked meat and the broth are used, together with other additives which may vary according to local consumption habits, for the preparation of tasty dishes.

Other types of dried meat, which are manufactured by a combination of drying with special treatments, are consumed raw, without rehydration and cooking. Some examples of this group of products are given below.

## Meat Drying in Combination With Additional Treatment

Meat drying after presalting, as described above, is the simplest and most efficient method of meat dehydration. Additional treatments used for some special dried meat products are curing, smoking and the utilization of spices and food additives.

Specific antimicrobial agents in smoke or spices or the antimicrobial properties of the curing substance, nitrite, may allow a less intensive dehydration of the meat. The resulting "semi-dry" products are in most cases consumed without rehydration, whereas rehydration is indispensable for common dried meat. In many countries, including developed countries, "semi-dry" products such as unsmoked and smoked raw hams (e.g. Parma ham, jamon serrano or smoked hams of the central European type), unsmoked or smoked dry sausages (e.g. salami, dry chorizo) or dried cured beef (Bündnerfleisch of

Switzerland) are not only popular because of the products' durability but particularly because they are delicious, high-quality meat specialities.

In developing countries, where the preservation aspect is even more important because of the lack of a cold chain, treatment carried out in addition to the drying of meat will be somewhat different and in some cases (e.g. intensive smoking over fire) the product quality is lowered rather than improved. The reasons for this additional treatment are in many cases adverse climatic or environmental conditions which do not allow the drying of meat without additional treatment. There are also of course other reasons for additional treatment, such as special flavours or special mixtures with non-meat ingredients, which may be preferred locally.

### Cured Dried Meat

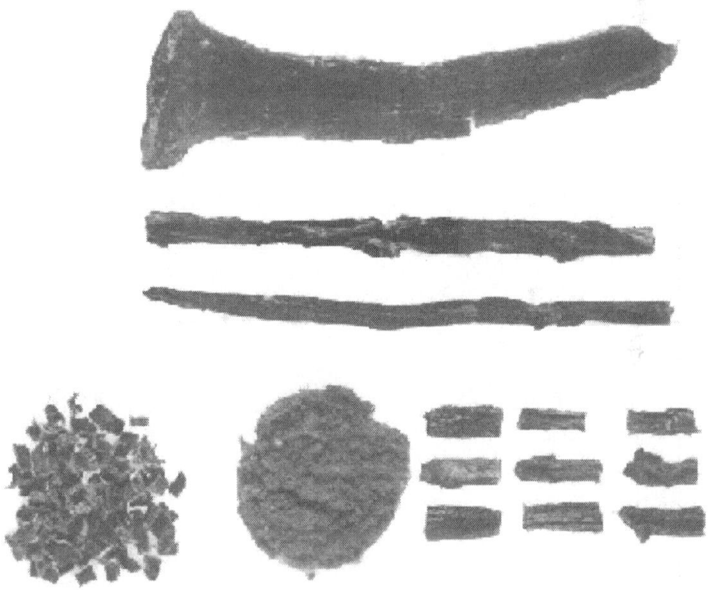

***Figure:*** *Whole strips and flat pieces of dried meat and dried meat comminuted to fragments of different sizes for preparing meals.*

Curing is the impact of nitrite on meat, in particular on the muscle pigment, myoglobin, which results in the formation of the pigment myochromogen and gives a stable red colour to muscle tissue. In addition, nitrite inhibits to some extent microbiological growth in the meat, but does so efficiently only in combination with low temperatures and/or low water activity. These effects are of particular importance for the shelf-life of raw hams and dry sausages and may also be of importance for non-intensively dried biltong, the South African dried meat, which may also be manufactured with nitrite or nitrate.

Apart from occasional use in biltong, it can be concluded that curing is not important in the manufacture of traditional dried meat products. The reasons are that a bright red colour is not desired in dried meat (because it will be rehydrated and used for cooking meals) and drying is generally so intensive that the inhibiting effect on microbiological growth is unnecessary. Curing substances must be handled very carefully as they are toxic even in low concentrations. Very small dosages are sufficient for the curing effect, about 200 ppm, that is, 2 g or less in 10 kg meat.

***Figure:*** *Preparing a meal of dried meat. As a first step the dried meat is put into boiling water.*

### Smoked Dried Meat

Smoking of meat is a technique in which meat is exposed directly to wood smoke which may be generated by a variety of methods. In smoke produced from wood there are various substances which contribute to the flavour and the appearance of the smoked meat product and which have a certain preserving effect on the product.

However, the preserving effect of common smoking is not very significant when storing the product without a cold chain. On the other hand, intensive or prolonged smoking may considerably increase the shelf-life of the product, but it also has an unfavourable effect on flavour. Whereas a light smoke aroma generally enhances the organoleptic properties of the product, intensive smoking has a negative influence on the quality, especially in the case of prolonged storage in which concentrated smoke compounds develop increasingly unpleasant tarry flavours.

In view of the above, smoking in order to preserve meat can only be considered as an emergency measure when no other preservation methods can be carried out. This may be the case during wet weather or generally under a humid climate, or when the preservation has to be completed as fast as possible because of the need of immediate transport, for instance after game-hunting.

Intensive meat smoking is always a combination of two effects, drying the meat by reducing its moisture content through hot air and the condensation of smoke particles on the meat surface together with their penetration into the inner layers of the product. Both have preservative effects and prolong the shelf-life of the product.

To smoke the meat, large strips and/or pieces, with and without bones, are dried by smoking in special drying/smoking places. The smoke is produced in these cases by glowing wood. Often, meat is prepared quickly by drying and smoking over a fire. In this case, the meat is not only smoked, but "half-cooked" or roasted. Normally, meat from this treatment is not well prepared and has to be consumed soon after drying, otherwise it will spoil quickly.

The quality of traditionally smoke-dried meat is generally poor. This is not only owing to poor meat quality or inadequate smoking devices, but mainly because smoke-drying is a rather rough treatment for the meat. The process is fast and has a certain preserving effect, but at the cost of quality.

Quality losses are even more obvious when failures in preparing the raw material occur. When, for example, the thickness of the meat parts to be smoked ranges from about 3 cm to 15 cm, uniform drying will not be achieved. The smaller pieces will be overdried and the thicker ones may still remain with a high moisture content in the product centre. The results of faulty drying and smoking are a too strong smoke flavour, lack of rehydration capacity of the smaller parts and fast spoilage of the thicker parts. For effective smoke-drying, the meat thickness should not exceed 7 cm to achieve products which are stable for a certain period without refrigeration.

Apart from primitive smoking places with just a fire below the meat, the construction of special smoking kilns has been suggested for smoke-drying of meat. The effect of light smoking could be of interest for the production of dried meat. Light smoking is not suitable for meat preservation without a cold chain, but it adds a smoke flavour to the product and inhibits the growth of moulds and yeasts on the product's surface owing to the fungistatic smoke compounds.

Thus light smoking may be used for the prevention of growth of moulds during the storage period of dried meat, especially under humid climatic conditions.

### Dried Meat with Spices and Additives

Various methods, typical for different regions, exist to produce this type of product. General guidelines for manufacture cannot be given because of the great variety of preparations, but the idea behind all of them is to combine the necessary preservation of meat with a typical flavour. In some cases the additives act as an absorbent with the aim of faster drying, and some spices may also act against bacterial growth.

## Typical Dried-meat Products With or Without Additional Treatment

Some examples of dried meat or dried and further processed meat manufactured in Africa, America and the Near East are described. No mention is made of the various dehydrated meat products well known in the Far East. These products are somewhat different owing to their sugar component. They will be the subject of future FAO studies.

### Odka

***Somalia and Other East African Countries:*** Odka is basically a sun-dried meat product made of lean beef and is of major importance to nomads in Somalia. In the face of perennial incidence of drought in the Horn of Africa, odka has become important since it is often prepared from drought-stricken livestock.

The production of odka is similar to the simple drying technique described earlier. However, the meat strips cut for drying are bigger and dry salting is usually applied instead of brine salting. After only four to six hours' sun-drying the large pieces of meat are cut into smaller strips and cooked in oil. After this heat treatment drying is continued and finally sauces and spices are added. For storage odka is again covered with oil and, when kept in a tightly closed container, it has a shelf-life of more than 12 months.

### Qwanta

***Ethiopia and Other East African Countries:*** Qwanta is manufactured from lean muscles of beef which are further sliced into long strips ranging from 20 to 40 cm and are hung over wire in the kitchen to dry for 24 to 36 hours. Prior to drying, the strips are coated with a sauce containing a mixture of salt (25 percent), hot pepper/ chilli (50 percent) and aromatic seasoning substances (25 percent).

After air drying the meat pieces may be further exposed to a light wood smoke and are then fried in butter fat and dried again to some extent. At this stage the product is ready for consumption or storage.

### Kilishi

***Nigeria and Other Arid or Semi-arid Zones of West Africa):*** Kilishi is a product obtained from sliced lean muscles of beef, goat meat or lamb and is made on a large scale under the hot and dry weather conditions prevailing from February to May. It is produced by sun-drying thin slices of meat.

However, recent experience indicates that kilishi can also be produced industrially using tray-drying in a warm air oven. Connective tissue and adhering fatty material are trimmed off the meat which is cut with a curved knife into thin slices of about 0.5 cm thickness, 15 cm length and as much as 6 cm width.

Traditionally, the slices of meat are spread on papyrus mats on elevated platforms or tables in the sun for drying. However, these papyrus mats may lead to hygienic problems, especially after repeated use. Therefore, easily washable corrosion-free wire nets or plastic nets are recommended for horizontal drying. The vertical drying method is also recommended in this case.

Sun-drying of kilishi could also be improved by the use of solar dryers. These devices will increase the rate of drying of the product and keep insects and dust from the product.

In the first stage of drying, which takes two to six hours, the moisture of the meat slices has to be reduced to about 40 to 50 percent. The slices are then put into an infusion containing defatted wet groundnut cake paste or soybean flour as the main component (about 50 percent), and is further composed of water (30 percent), garlic (10 percent), bouillon cubes (5 percent), salt (2 percent) and spices such as pepper, ginger and onion. The "dried" slices of meat should absorb the infusion up to almost three times their weight.

After infusion, the wet product is again exposed to the sun to dry. Drying at this stage is much faster than at the first stage. When the moisture content of the slices has been reduced to 20 to 30 percent, a process which takes two to three hours depending on weather conditions and the dimensions of the product, the slices are finally roasted over a glowing fire for about five minutes.

The roasting process helps to enhance desirable flavour development and to inactivate contaminating micro-organisms. Roasted kilishi is therefore superior in flavour to the unroasted version.

After roasting, the final moisture content ranges between 10 to 12 percent. It will decrease during storage at room temperature to as low a level as 7 percent. When packaged in hermetically sealed, low density plastic bags the product remains remarkably stable at room temperature for a period of about one year.

### Biltong

***Southern African Countries:*** Biltong is a well-known salted, dried meat prepared from beef or antilope meat. Most muscles in the carcass may be used but the largest are the most suitable. The finest biltong with the best flavour is made from the sirloin strip and the most tender is derived from the fillet.

The meat is cut into long strips (1 to 2 cm thick) and placed in brine, or dry-salted, which is actually the most popular method. Common salt, preferably coarse salt (1 to 2 kg for 50 kg of meat), or salt and pepper are the principal ingredients used, although other ingredients such as sugar, coriander, aniseed, garlic or other spices are included in some mixtures to improve flavour. In most cases nitrate or nitrite is added to achieve a red colour and the typical flavour of cured meat.

The addition of 0.1 percent potassium sorbate to the raw meat is permitted in South Africa as a preservative. The salt/spice mixture is rubbed into the meat by hand and the salted strips are then transferred to a suitable container. It is recommended that a little vinegar be sprinkled on each firmly packed layer in the container.

Biltong is left in the curing brine for several hours, but not longer than 12 hours (otherwise it will be too salty), and then dipped into a mixture of hot water and vinegar (approx. 10:1). The biltong is now ready for sun-drying for one day. Then the strips are moved into the shade for the rest of the drying period. The product is usually not smoked, but if it is smoked only light cold smoking is recommended, which takes one to two weeks under sufficient air circulation. The biltong is ready when the inside is soft, moist and red in colour, with a hard brown outer layer.

Biltong is sold in sticks or slices. The usual shelf-life is several months without refrigeration and packaging, but in airtight packages the product stores well for more than one year. Biltong is not heated during processing or before consumption. It is eaten raw and considered a delicacy.

### Pastirma

***Turkey, Egypt, Armenia:*** Pastirma is salted and dried beef from not too young animals. In some areas camel meat is also used. The meat is taken from the hindquarters and is cut into 50 to 60 cm long strips with a diametre of not more than 5 cm. The strips are rubbed and covered with salt and nitrate. The dosage of the nitrate in relation to the meat is 0.02 percent, that means 2 g of nitrate for 10 kg of meat. Several incisions are made in the meat to facilitate salt penetration.

The salted meat strips are arranged in piles about 1 m high and kept for one day at room temperature. They are turned over, salted again, and stored in piles for another day. Thereafter the meat strips are washed and air-dried for two to three days in summer and for 15 to 20 days in winter. After drying the strips are piled up again to a height of 30 cm and pressed with heavy weights (approx. 1 tonne) for 12 hours. After another drying period of two to three days the meat pieces are again pressed for 12 hours. Finally the meat is again air-dried for 5 to 10 days.

After the salting and drying process, the entire surface of the meat is covered with a layer (3 to 5 mm thick) of a paste called cemen, which consists of 35 percent freshly ground garlic, 20 percent helba (i.e. ground trefoil seed), 6 percent hot red paprika, 2 percent mustard, and 37 percent water. Helba is used as a binder of the paste; the other ingredients are spices, but garlic is the most important as it is antimycotic.

The meat strips covered with cemen are stored in piles for one day, and thereafter are dried for 5 to 12 days in a room with good air ventilation, after which the pastirma is ready for sale. Thus, the production of pastirma requires several weeks. However, not much energy is required since most of the salting and drying is done at room temperature. The final product has an average water activity ($a_w$) of 0.88. The $a_w$ should not fall below 0.85 or the meat will be too dry. The salt content should range between 4.5 and 6.0 percent.

The product is mould-free for months at ambient temperature even in summer. Pastirma thus has a better microbiological stability than biltong.

### Charque

***Brazil and Other South American Countries:*** Charque consists of flat pieces of beef preserved by salting and drying. The fresh, raw meat from the fore- and hindquarters is cut into large pieces of about 5 kg, which should not be more than 5 cm thick. The pieces are submerged in a saturated salt solution for about one hour in barrels or cement vats. On removal from the brine, the meat is laid on slats or racks above the brine tank to drain.

For dry-salting, the flat meat pieces are piled on a sloping, grooved, concrete floor under a roof. To form a pile, salt is spread evenly over the floor about 1 cm high. Then a layer of meat is put on the salt. The meat is covered with another (1 cm) layer of salt followed by adding another layer of meat, and so on until the alternate layers of salt and meat reach a height of about 1 m. The pile is then covered with a few wooden planks and pressed with heavy stones.

After eight hours the pile is restacked so that the top meat goes to the bottom of the pile. The restacking process with fresh layers of salt is repeated every day for five days.

The salted meat is then ready for drying. Before initiating drying, the meat pieces are subjected to rapid washing to remove excess salt adhering to the surface. The meat pieces may also be passed through a pair of wooden rollers or a special press to squeeze out some surplus moisture and flatten the meat slabs. The meat is then spread out on bamboo slats or loosely woven fibre mats in a shed or, in industrial production, exposed to the sun on wooden rails which are oriented north-south, thus permitting an even solar coverage.

Initial drying, directly in the sun, is limited to a maximum period of four to six hours. This period of exposure may be subsequently lengthened to a maximum eight hours. Temperatures in excess of 40°C on the meat surface should be avoided. To ensure even drying over the extended muscle pieces, the meat is placed on the rails during the morning and removed again in the afternoon. The meat pieces are exposed to the sun each day over a period of four to five days. After each period of exposure the pieces are collected, stacked in piles on concrete slabs and covered with an impermeable cloth to protect them against rain and wind and to hold the heat absorbed.

When sufficiently dry, the meat pieces are either sold without prior packaging or wrapped in jute sacks. Plastic sacks are not suitable, because the product still contains a certain proportion of its original moisture content, and this moisture must be allowed to drain freely from the product.

Charque keeps for months under ambient room conditions and is resistant to infestation by insects and growth of moulds.

# 2

# Meat Preservation by Thermal Treatment

## Characteristics of Heat-treated Preserved Meat and Meat Products

The prolonged shelf-life of heat-treated meat and meat products is achieved through reducing growth of, or inactivating, micro-organisms by a thermal process. The principal steps of the heat preservation method are to:

- place the product in a container (can, glass jar, pouches of synthetic material or laminate with aluminium) which is hermetically sealed after filling and which is impermeable to any external substances; and
- submit the hermetically sealed product to thermal treatment with a defined temperature and time combination.

## Equipment For Thermal Treatment

Thermal or heat treatment is done by submerging the products in cooking vats or pressure cookers which contain hot water or steam or a mixture of both. It can be performed under pressure in pressure cookers (retorts, autoclaves) in order to reach temperatures above 100°C ("sterilization"). Sterilization is the most important and efficient type of heat treatment, since foods free from viable micro-organisms can be obtained and most of these products can then be stored without refrigeration. In contrast, temperatures up to 100°C can be achieved in simple cooking vats ("pasteurization"). A certain amount of micro-organisms resist this moderate heat treatment and the resulting pasteurized products must consequently be stored under controlled temperatures.

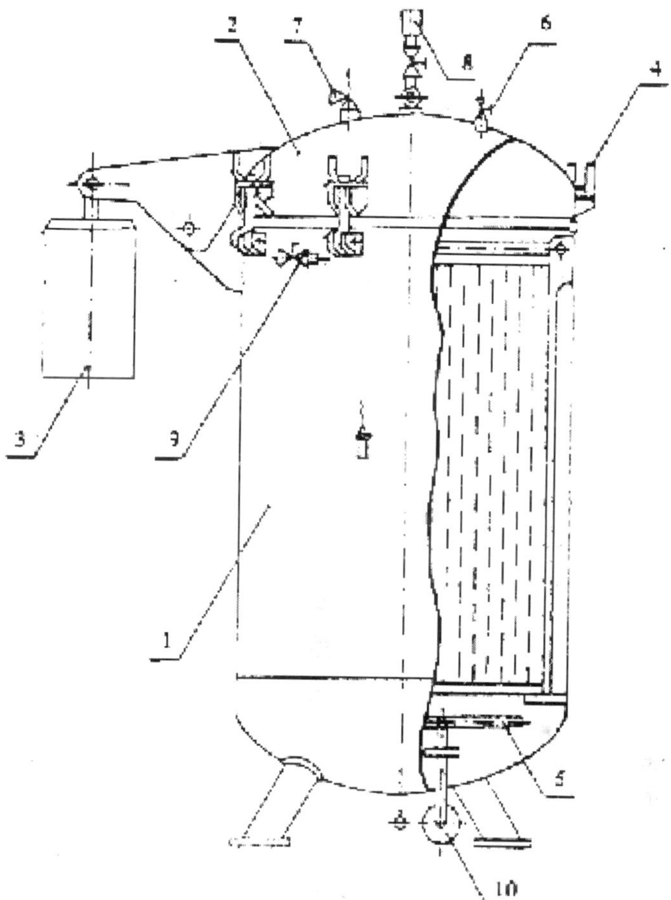

***Figure:*** *Retort cooker. 1. body, 2. lid, 3. counterbalance, 4. nuts, 5. heater, 6. vent, 7. relief valve, 8. pressure gauge, 9. water supplyvalve 10. steam supply valve.*

In simple retort cookers (autoclaves) pressure is generated either by a direct steam injection, by heating water up to temperatures over 100° C or by combined steam and water heating. The retort must be fitted with a thermometre, a pressure gauge and a relief valve. Modern autoclaves may also have revolving drums, speeding up the heating of the products.

After thermal treatment the product must be chilled as quickly as possible, in order to avoid overcooking. Hence, this operation is done within the cooker by introducing cold water. The contact of cold water with steam causes the latter to condense with a rapid pressure drop in the retort. However, overpressure simultaneously built up during thermal treatment within the cans, jars or pouches remains for a certain period, and may induce permanent deformation or damage of these containers.

Therefore, a high pressure difference between the cooker and the internal pressure in the containers must be avoided. This is generally achieved by a blast of compressed air into the retort or by sufficient hydrostatic pressure of the introduced chilling water.

### Containers For Thermally Treated Preserves

Containers for heat-preserved food must be airtight in order to avoid recontamination by environmental microflora. Moreover, no traces of undesirable substances which the packaging material may contain, such as heavy metals (lead, tin), should be permitted to migrate into the product. Currently, most of the thermally preserved products are in metal containers (cans), others are packed in glass jars or plastic or aluminium/plastic laminated pouches.

Metal containers are *cans* (tins) produced from *tinplate* or *tin-free* steel. They are usually cylindrical. However other shapes such as rectangular or pear-shaped cans are also encountered. Tinplate consists of steel plate, plated with tin on both sides. The steel body usually is 0.22 to 0.28 mm in thickness. The tin layer is from 0.385 to 3.08 µm. Tin-free steel plates have other protective coatings such as chromium, aluminium, or nickel, which are generally even thinner than the tin layers of the tinplate.

The cans (tins) usually consist of three elements, i.e. the body and two ends. The seaming panel of the ends is fitted with a synthetic lining. The ends are fastened to the body with seams made by a seaming (closing) machine. Proper seaming is vital to the tightness of the can. Any leak causes recontamination, in particular during chilling. This will result in swelling of the cans during storage, and creates a risk of food poisoning.

For smaller and easy-to-open cans aluminium is frequently used. *Aluminium cans* are deep-drawn, i.e. the body and the bottom end are formed out of one piece and only the top end is seamed on after the filling operation. The advantages of aluminium cans are low weight, resistance to corrosion, good thermal conductivity and recyclability, but these cans cannot be soldered or welded. They are less rigid and more expensive than steel plate.

*Glass jars* are used less often for meat products because of their fragility. They consist of a glass body, and a metal lid. In households, glass jars with glass lids are often used. The seaming panel of the metal lid has a lining of synthetic material. Glass lids are fitted by means of a rubber ring.

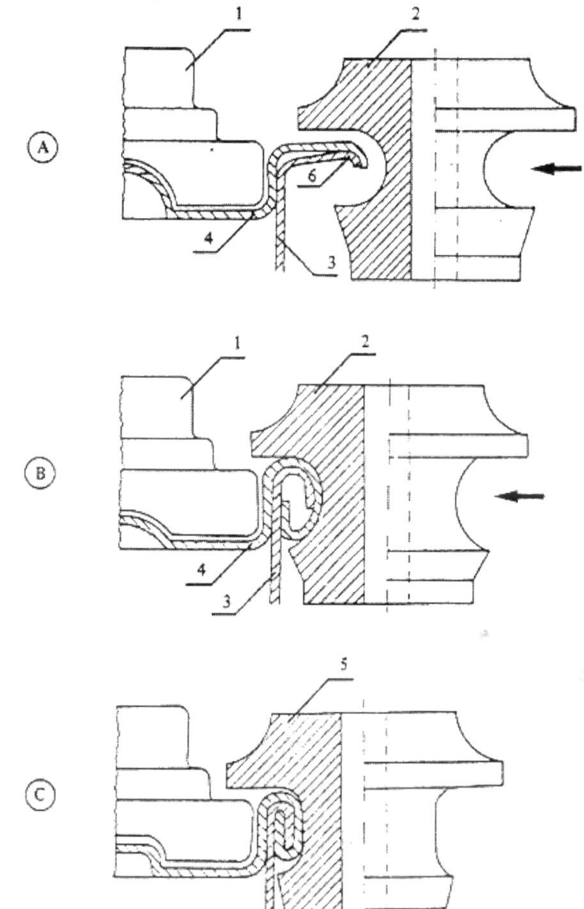


***Figure:*** *Principles of canseaming operation. (A) initial stage of curling; (B) fully developed curling; (C) seam tightening. 1. seaming chuck, 2. seaming roll, 3. tin wall, 4. tin end, 5. seaming roll, 6. lining compound.*

Containers made either of *synthetic material* or *laminates of aluminium foil with synthetic material* are of growing importance in thermal preservation. Heat-resistant plastic pouches, which are closed by clip, are usually made of polyester (PETP) and used for frankfurters in brine or ready-to-eat dishes. From laminated films, for instance, polyester/polyethylene (PETP/ PE) or polyamide/polyethylene (PA/ PE), relatively rigid containers can be made, usually by deep drawing, which are used for filling with pieces of cured ham or other kinds of prepared meat. Widely used for small portions, particularly of sausage mix, are round containers formed out of a laminate of aluminium foil and polyethylene (PE) or polypropylene (PP). PE or PP permit the heat-sealing of these containers, which can then even be subjected to intensive heat treatment.

### *Meat Products Suitable For Canning*

Basically all meat products which require heat treatment to prepare them for consumption are also suitable for heat preservation. Only meat products which do not receive any form of heat treatment before being consumed, such as dried meat, raw hams or dry sausages, are naturally not suitable for canning. These products are conserved by a low pH value and/or low water activity.

The following groups of meat products, when not consumed freshly cooked, are frequently found as canned products:

- cooked ham
- sausages with brine of the frankfurter type
- sausage mix of the bologna or liver sausage type
- meat preparations such as corned beef, chopped pork, etc.
- ready-to-eat dishes with meat ingredients such as beef in gravy, chicken with rice, etc.
- soups with meat ingredients such as chicken soup, oxtail soup, etc.

### *Organoleptic, Physical and Microbiological Aspects of Thermal Treatment*

The intensity of heat treatment has not only a decisive impact on the inactivation of micro-organisms, but also on the organoleptic quality of the product. There are products which undergo intensive temperature treatment without significant losses in quality. On the other hand, other products may deteriorate considerably in taste and consistency after sterilization. In these cases less intensive thermal treatment is required but, at the same time, other hurdles, such as low pH value and/or water activity or a lower storage temperature, have to be built up in order to inhibit bacterial growth.

The intensity of thermal treatment can be defined in physical terms. The term widely used under practical conditions is the F-value, with which the lethal effect of heat on micro-organisms can be defined. The thermal death time for different micro-organisms calculated at 121°C and expressed in minutes, is used as the reference value.

The thermal death time for spores of *Clostridium botulinum* at 121°C is 2.45 minutes or in other words, an F-value of 2.45 is needed to inactivate all these spores in the product at 121°C. Spores of other micro-organisms are more or less heat resistant. Vegetative cells of micro-organisms are generally destroyed at temperatures of less than 100°C and therefore play no role in the F-value calculations. The definition of the F-value at 121°C is as follows:

F = 1: lethal effect at 121°C on micro-organisms after 1 minute

F = 2(3, 4, etc.): lethal effect at 121°C on micro-organisms after

2(3, 4, etc.) minutes. In Tables 2 and 3 some examples are given for F-values obtained at different time/temperature combinations:

***Table:*** *F-values corresponding to various temperatures*

| | | |
|---|---|---|
| 95°C | per minute: | F = 0.003 |
| 100°C | per minute: | F = 0.008 |
| 105°C | per minute: | F = 0.025 |
| 110°C | per minute: | F = 0.079 |
| 115°C | per minute: | F = 0.251 |
| 121°C | per minute: | F = 1.0 |
| 125°C | per minute: | F = 2.51 |
| 130°C | per minute: | F = 7.94 |

***Table:*** *F-values in relation to temperature and time*

| | |
|---|---|
| *In order to reach F = 1, the following time-temperature combinations are required:* | |
| 110°C during 12.5 minutes | or |
| 116°C during 3 minutes | or |
| 121°C during 1 minute | or |
| 130°C during 0.13 minutes | |
| In order to reach F = 4, the following parametres are required: | |
| 110°C during 50 minutes | or |
| 116°C during 12 minutes | or |
| 121°C during 4 minutes | or |

*Contd...*

130°C during 0.5 minutes

In order to reach F = 0.6, the following parametres are required:

110°C during 7.5 minutes or

116°C during 2 minutes or

121°C during 0.6 minutes or

130°C during 0.08 minutes

The lethal effect can be shown in the reduction (in percentage) of the total number of micro-organisms present in the product. The destruction of micro-organisms is at an exponential rate, which means that the higher the initial bacterial load (using the same time-temperature combination), the higher the number of surviving bacteria.

***Table:*** *Decimal reduction rates during heat treatment*

| | ***Remaining micro-organisms*** | | |
|---|---|---|---|
| ***Initial bacterial load (micro-organisms/g)*** | ***Destruction rate 90%*** | ***Destruction rate 99%*** | ***Destruction rate 99.9%*** |
| 10 million | 1 million | 100 000 | 10 000 |
| 1 million | 100 000 | 10 000 | 1 000 |
| 100 000 | 10 000 | 1 000 | 100 |
| 10 000 | 1 000 | 100 | 10 |
| 1 000 | 100 | 10 | 1 |

The initial bacterial load and the destruction rate are shown in table above.

Table demonstrates the importance of proper meat hygiene. Highly contaminated raw material with bacterial loads of 10 million per g will, even after intensive heat treatment, still give final products with a rather limited shelf-life because of the high remaining rate of contamination.

Since the heat treatments will in many cases not be intense enough to destroy all spores, it is important that cans be chilled as rapidly as possible after retorting and that storage temperatures generally not exceed 20 to 25°C.

The nature of the heat-preserved product, its pH, amount of salt and other curing agents, and the number of spores present, together with retorting time and temperature, determine the degree of commercial sterility and product safety. It has been shown that F-values of 4 in heat-preserved products will guarantee commercial sterility. Products with F-values below this level need additional measures such as lowering the pH or the $a_w$ or refrigerated storage for their microbiological safety.

Micro-organisms have two adverse effects in improperly treated heat-preserved products:

- organoleptic deterioration through protein degradation;
- food poisoning by bacteria and/or toxins.

The food-poisoning aspects require special care during production and storage of heat-treated preserves, bearing in mind that some heat-resistant micro-organisms are able to produce dangerous toxins, amongst them *Clostridium botulinum*, which may have fatal consequences.

### Practical Application of F-values

By measuring the temperature of the product periodically during thermal treatment, the final F-value can be determined. It is obvious that during thermal treatment the product temperature will rise constantly. The temperature taken in the centre of the container after each minute of heat treatment corresponds to a certain F-value. These partial F-values are added up (for example by using special tables containing F-values corresponding to temperatures from 90°C to 140°C) and the sum is the overall F-value of the product.

The exact F-value is of special importance for the producer because:

- it ensures appropriate thermal treatment of the product, thus avoiding over-or undercooking;
- it enables the product's storage time to be determined.

In practice it is not necessary to calculate the F-value repeatedly for the same type of batch processed in the cannery. The F-value can be determined once for each batch according to the size of the containers and intensity and duration of thermal treatment. If these parametres remain unchanged, the F-values will not be subject to alteration.

### Categories of Heat-treated Preserves

***Pasteurized Products:*** Only slight thermal treatment. Temperatures reached in the product centre are in the range of 82°C and below 100°C ("pasteurization"). The F-value cannot be determined, remaining almost at zero.

***Inactivated:*** most vegetative micro-organisms

***Not Inactivated:*** spores of *Bacillus* and *Clostridium*

***Storage Required:*** uninterrupted cold chain (2–4°C), up to six months

***Cooked Preserves:*** Thermal treatment only with boiling water (no pressure cooker).

Temperature reached in the product centre is up to 100°C. Low F-value.

***Inactivated:*** all vegetative micro-organisms

***Not Inactivated:*** spores of *Bacillus* and *Clostridium*

***Storage Required:*** not higher than 10°C for one year. Spores will not grow under these conditions.

***"Three-quarter" Preserves:*** Thermal treatment in pressure cooker. Temperatures reached in the product centre are between 108 and 112°C. F-value 0.6 to 0.8.

***Inactivated:*** all vegetative micro-organisms, spores of *Bacillus*

***Not Inactivated:*** spores of *Clostridium*

***Storage Required:*** not higher than 15°C for one year. Spores of *Clostridium* will not grow under these conditions.

"Full" preserves stable under temperate conditions:

Intensive thermal treatment in pressure cooker. Temperature reached in the product centre is about 121°C. F-value 4 to 6 ("sterilized product").

***Inactivated:*** all micro-organisms except thermophilic spores

***Storage Required:*** ambient temperature (for one year), but not tropical conditions (40°C or more).

"Full" preserves stable under tropical conditions:

Very intensive thermal treatment, with a long period of 121°C or higher in the product centre. F-value of 12 and more.

***Inactivated:*** all micro-organisms including thermophilic spores

***Storage Required:*** ambient temperature even under tropical conditions (up to four years).

***Shelf-stable Preserves:*** This group of preserves is different from those mentioned previously, since preservation is achieved not only by thermal treatment, but also by utilizing other means to prevent microbiological growth such as nitrite, low water activity and/or low pH. This combined effect has the advantage of a fully shelf-stable product under all ambiental conditions without undergoing intensive thermal treatment (less than 100°C) and without major losses in organoleptic quality.

Shelf-stable preserves are a fairly new development in the food sector and will certainly gain special importance in countries without an uninterrupted cold chain.

Thermal treatment of different intensity for different products is used to avoid deterioration, which varies from product to product. The following table gives some examples on how thermal treatment should be conducted. It is a general rule in this context that products in smaller containers can undergo more intensive thermal treatment because of faster heat penetration.

## Impact of Packaging Methods on Meat Preservation

In most developing countries meat carcasses are handled and stored either without or with only minimal refrigeration. A fast-turnover system is used, ensuring that meat slaughtered during the night and morning is sold and consumed the same day. This well-proven traditional method functions satisfactorily because in the short period between slaughtering and consumption, micro-organisms cannot increase to the extent of spoiling the meat and making it inedible. This system is eminently suitable for rural villages and small towns in many developing countries, particularly because local cooking habits do not call for ripened and tender meat.

This fast turnover of meat normally does not lead to meat spoilage, although massive contamination during slaughtering and meat handling frequently occurs. Fortunately, in most cases toxic micro-organisms do not develop to such an extent to threaten consumers' health, provided that the time between slaughter and meat cooking is short and cooking is done properly and with sufficiently high temperatures.

In traditional meat handling, fresh meat is generally not packaged at all or just wrapped in paper, leaves, etc. Meat, traditionally preserved by drying, is sometimes packaged in linen bags, baskets or pottery to facilitate storage and transport and to provide some kind of protection against dirt, insects, etc.

With higher concentrations of population, however, this traditional system now becomes outmoded in some places in developing countries because more time is needed between slaughtering and ultimate consumption. Meat frequently has to be stored, transported, prepared and distributed through a retailer or supermarket, all of which is considerably time-consuming. In order to safeguard fresh meat during this extended time, certain methods of preservation have to be applied. Refrigeration is the obvious solution, but this is expensive and therefore frequently not available in developing countries. Energy-saving storage methods are therefore particularly relevant in underdeveloped areas. For both methods, either using the refrigeration technology or energy-saving methods to extend the shelf-life of meat and meat products, proper packaging has an important part to play.

### *Purpose of Packaging Fresh and Processed Meat*

The purpose of packaging is primarily to protect foodstuffs during the distribution process, including storage and transport, from contamination by dirt, micro-organisms, moulds, yeasts, parasites, toxic substances or those influences affecting smell and taste or causing loss of moisture. Packaging should help to prevent spoilage, weight losses and enhance customer acceptability.

Simple packaging without further treatments is less effective in prolonging the shelf-life of meat and meat products. Frequently full advantage of packaging can only be achieved in combination with preservation methods.

### *Factors Affecting the Shelf-life of Meat and Meat Products*

Though meat handling, storage and consumption may differ from one place to another, the factors limiting the shelf-life of these products are the same.

There are *endogenous* factors, such as:

- pH-value or the degree of acidity of the product;
- $a_w$ value or the amount of moisture available in the product; and *exogenous* factors, such as:
- oxygen (from the air);
- micro-organisms;
- temperature;
- light; and
- evaporation and desiccation.

## Combined Effect of Ph- and $A_w$-value

Generally speaking the shelf-life of meat and meat products will be longer the lower the pH-value and/or $a_w$-value. Both factors (either pH or $a_w$ alone or the two together) have a decisive influence on the growth of micro-organisms in food. However, there are limits for most meat products regarding decreased pH-value and $a_w$-value, particularly for organoleptic reasons. Except for some special products, consumers do not want meat products to be excessively acidic or dry.

Uncanned meat products can be classified into three storage groups according to their pH and $a_w$. Each group requires different storage conditions.

*Highly perishable* meat products have a pH-value above 5.2 and an $a_w$-value above 0.95; refrigeration at or below +5°C is needed. These are raw fresh meat (without additives), bologna-type sausages, cooked sausages and cooked ham.

*Perishable* meat products have a pH-value below 5.2 or an $a_w$-value below 0.95. Refrigeration at or below +10°C is needed to keep them stable. Products such as meat or poultry pieces in vinegar jelly (acid) and semi-dry sausages or hams belong to this group.

*Shelf-stable* products have a pH-value of or below 5.2 and an $a_w$-value of or below 0.95, or only a pH-value below 5.0, or only an $a_w$-value below 0.91. No refrigeration is required in these cases, the products remaining stable under ambient temperatures. The most common products in this group are the various kinds of dried meat.

Under the above conditions no microbial growth in meat and meat products will occur. However, this does not mean that the products remain stable for an undetermined period. Their shelf-life will be limited by chemical or physical deterioration, by rancidity and discolouration.

In this situation the product quality will benefit from the application of suitable packaging materials, which reduce the physical and chemical influences on the product or protect the product completely. The following noxious influences may occur.

### Oxygen

The oxygen content in the air is about 20 percent. If oxygen affects meat and meat products during prolonged storage periods, it will change the red colour into grey or green and cause oxidation and rancidity of fats with undesirable off-flavours.

The foils used for food packaging differ in their permeability to oxygen. The lower the oxygen permeability of the packaging material, the more efficient will be the protection of product quality. The best protection will be achieved using oxygen-proof packaging films together with vacuum packaging of the product. This ensures that practically no oxygen is left in the package and no oxygen will penetrate from the air into the product.

### Light

The prolonged exposure of meat and meat products to daylight or artificial light accelerates oxidation and rancidity because light provides the energy for these processes.

Transparent packaging films give no protection against light influences. Therefore, for products under strong light exposure, coloured or opaque films should be preferred. Films laminated with aluminium foil are absolutely impermeable to light. Products in transparent packaging film are sufficiently protected when kept in the dark or under moderate illumination.

### Evaporation

Fresh foods with a relatively high moisture content such as meat, fresh sausages, cooked ham, etc. will have considerable losses of weight and quality by evaporation during storage if they are not packed. The packaging material must therefore be sufficiently vapour-proof. Most plastic films used for food packaging comply with this requirement.

### Secondary Contamination

During slaughtering, carcass dressing, meat cutting and/or processing, the contamination of meat to some extent cannot be avoided. The further growth of micro-organisms in meat and meat products cannot be stopped through packaging only.

However, secondary contamination of these foods, for example by contact with dust, dirty surfaces and hands, can definitely be prevented through proper packaging, preferably with plastic films which are absolutely impervious to agents causing secondary contamination.

### Suitable Materials and Equipment For Packaging Meat and Meat Products

Packaging films can be subdivided into *cellulose films, plastic films* and *aluminium foil.* They can either be used as monofilms or as two or more different films laminated together. These materials differ in:

- oxygen permeability;
- water vapour barrier;

- resistance to hot and cold temperatures; and
- mechanical strength.

Nearly all available films are of thermoplastic materials, and therefore *heat sealable*, resulting in hermetically sealed plastic pouches, bags, etc.

A high oxygen barrier is important in the application of films for packaging meat and meat products. Films made of polyvinylchloride (PVC), polyethylene (PE) or polypropylene (PP) have a relatively high oxygen permeability, whereas polyvinylidenchloride (PVDC), polyester (PETP), polyamide (PA) and cellulose film (ZG) are less or almost non-permeable to oxygen. The latter materials are therefore better suited for packaging meat and meat products. However, the materials of the first group are frequently used as laminates with materials of the second group in order to achieve special effects regarding mechanical strength, heat sealing properties or making the package practically impermeable to both oxygen and water vapour.

For the efficient utilization of these materials, air must be drawn out completely from the packages with the meat product ("vacuum-packaging") and the package should be hermetically closed by heat sealing or metal clip. Vacuum-packaging will inhibit the growth of yeasts, moulds, and most aerobic bacteria under refrigeration temperatures. However, facultative anaerobes such as acid-producing bacteria grow freely under adequate $a_w$-conditions. These micro-organisms do not threaten the health of consumers but may affect the quality of the products by negatively influencing colour and taste. Therefore efforts should be made to keep the initial number of bacteria low, which can be obtained by simple hygienic measures during cutting and packaging, such as frequent handwashing, clean tools and clean tables.

The above general considerations refer to vacuum-packaged meat and meat products with a high moisture content. The situation is different for the traditionally preserved meat and meat products described. All these products are shelf-stable because of their low water activity and therefore do not require refrigeration. They are partly dried in the sun, or others are dried and smoked over a wood fire.

These low-moisture products are hygroscopic and should be protected in the case of high air humidity. Traditional wrapping methods using paper, jute or linen will not inhibit the penetration of water vapour. However, a simple polyethylene bag is sufficient as a water vapour barrier, but care must be taken that the bag is tightly closed. If plastic bags are used for packing, then the pieces of dry meat should be cut to a certain length so that they can be better arranged in the bags. The bag should be bound with rope, rubber or cellotape. Bags stuffed with meat in this way are best stacked into a carton.

Longer storage times for traditionally preserved dried meat can be achieved with vacuum-packaging in laminated films, which protect against humidity and oxidation by oxygen from the atmosphere. Vacuum-packaging can also gain importance for traditionally preserved products. Although they are shelf-stable, exposure of these products to oxygen and the humidity of the surrounding atmosphere will cause unfavourable alterations (rancidity, loss of flavour, growth of certain micro-organisms) over the long term. Air and vapour-tight packaging will ensure a prolonged shelf-life.

Simple vacuum-packaging machines, suited to less developed regions, are available. They consist of a chamber with a removable lid, so that the plastic pouch or bag containing the product can be placed inside the chamber, which is evacuated with an electrically driven vacuum pump. By closing the lid, air can be drawn out of the chamber and also out of the open bag. While still in the vacuum chamber, the bag is hermetically closed by heat sealing by means of a manual or automatic electrically heated sealing device. After being sealed the bag can be exposed to ambiental conditions without air penetrating the package.

3

# Basic Methods of Quality Control

The quality of meat and meat products is defined by the following criteria:

- palatability (typical texture and consistency, juiciness, good flavour);
- proportion of lean meat to fat;
- freshness and adequate conservability of the products;
- absence of harmful micro-organisms or substances; and
- appropriate (preferably minimal) use of additives and meat extenders.

The different criteria need different methods of quality control, such as:

- organoleptic evaluation
- physical test methods
- chemical analysis
- microbiological examination

According to the accuracy needed, the control method applied can be simple or more complicated and different auxiliary technical devices must be used. In order to inform consumers and meat processors about the quality of meat and meat products, simple and fast control methods are best suited in many cases, although exact details on residues, toxins and special food components can only be obtained through specialized laboratories.

Basic methods for quality control must involve little or no equipment and obviously sensory evaluation will be most important.

Some physical tests, however, can easily be performed using simple instruments such as thermometres, manometres, scales, etc. By contrast, chemical and microbiological tests are more complicated. These methods not only require standard equipment but also skilled and experienced personnel to do the tests and to interpret the results.

The following mainly refers to the basic methods of quality control used in connection with the handling and processing of meat. These control methods can easily be applied for meat products processed with simple meat preservation techniques.

## Organoleptic Evaluation

Organoleptic evaluation consists in describing the attributes of food, in this special case of meat and meat products, that can be perceived by the sense organs. The attributes to be evaluated are appearance, colour, texture and consistency, smell and taste.

### *Appearance*

The way meat looks, either as a carcass or as boneless meat cuts, has an important impact on its objective or subjective evaluation. Grading is an objective evaluation method in this context. Traditional methods of carcass grading after slaughter involve the aspect of beef or pork sides, poultry carcasses, etc. Skilled graders are able to classify different carcasses by checking the size, the volume of muscular tissue, fat layers, etc.

Although in modern grading procedures more and more technical equipment has been incorporated, visual methods are still in use. They can be of special value in most developing countries where no extremely sophisticated methods are needed.

The way the consumers or the processors check the appearance of meat is subjective. Differences will be registered in the relation of lean meat and fat including the degree of marbling or in the relation of bones and lean meat. Furthermore, unfavourable influences can be detected such as unclean meat surfaces, surfaces too wet or too dry, or unattractive blood splashes on muscle tissue.

Processed meat, on the other hand, can roughly be evaluated by its appearance according to the different raw materials of which the product is composed and where the use of some components is exaggerated (for instance too many particles of visible fat or connective tissue, etc.). Special product treatments (for instance chilling, freezing, cooking, curing, smoking, drying) or the kind and quality of portioning and packaging (casings, plastic bags, cans) will be recognized by evaluating the appearance.

### Colour

Under normal circumstances the colour of meat is in the range of red and may differ from dark red, bright red to slightly red; but also pink, grey and brown colours may occur. In many cases the colour indicates the type and stage of the treatment to which the meat has been subjected, as well as the stage of freshness.

In judging meat colour, some experience is needed to be able to distinguish between the colour which is typical for a specific treatment or which is typical for specific freshness. Furthermore, meat deriving from different species of animals may have rather different colours, as can easily be seen when comparing beef, pork and poultry meat.

The natural colour of fresh meat, except poultry meat, is dark red, caused by the muscle pigment, myoglobine. Fresh meat surfaces which have been in contact with the air for only a short period turn into a bright red colour because of the influence of the oxygen in the air. Oxygen is easily aggregated to the myoglobine and drastically changes the colour of the meat surfaces exposed to it. On the other hand, in the absence of oxygen, for example in meat cuts packaged in impermeable plastic bags, meat surfaces remain or become dark red again. The same conditions generally prevail in the interior of meat cuts which are not reached by oxygen. Changes from dark red to bright red are therefore typical and are normal reactions of fresh meat.

Meat which is in the process of losing its freshness, however, no longer shows a bright red colour, even when intensively exposed to the air, because of the partial destruction of the red meat pigment which results in a grey, brown or greenish colour. Once these conditions occur the consumer has to decide, after carefully checking the appearance, together with testing smell and taste, whether the meat has to be discarded as a whole or whether use can be made of some parts which so far have not been altered.

Remarkable changes in the meat colour occur when fresh meat has been boiled or cooked. It loses its red colour almost entirely and turns to grey or brown. The reason for this is the destruction of the myoglobine through heat treatment. On the other hand, it has long been known that after pickling (curing) fresh meat with curing ingredients (nitrite), the meat colour remains red during longer storage periods, after ripening, drying and even after intensive heat treatment. Obviously the original meat colour has not been conserved, but a chemical reaction has taken place during the curing process transforming the unstable pigment of the fresh meat into a stable red pigment. This is the typical colour shown in sausages of all types, raw and cooked hams, corned beef, etc.

It should also be noted that cured products have a longer shelf-life than fresh meat because of the conserving effect of the curing salt. However, cured products will also deteriorate under unfavourable conditions, cooked cured products sooner than raw cured products. Cured products with a decreasing keeping quality can be recognized when the red colour becomes pale or changes to grey or green.

## Texture and Consistency (Tenderness and Juiciness)

Meat prepared for the consumer should be tender and juicy. Meat tenderness depends on the animal species from which the meat originates. Lamb, pork and poultry meat are sufficiently tender after slaughter, but beef requires a certain period of maturation to achieve optimal eating quality. Texture and consistency, including juiciness, are an important criterion, still neglected by many consumers, for the eating quality of meat. Often consumers do not know that the eating quality of meat can be upgraded by ripening, especially in the case of beef and similar meats. There is also a great deal of consumer negligence in how to prepare meat. It should be cooked to become sufficiently tender, but cooking should not be too intense otherwise the meat becomes dry, hard and with no juiciness.

The simple way to check the consistency of foods is by chewing. Although this test seems easy, in practice it is rather complicated. Taste panelists need experience, particularly when the different samples have to be ranked, for example which sample is the toughest, the second toughest or the most tender.

The texture is of less importance in meat products, such as cured or canned products, sausages, etc., because they are either made of comminuted meat and/or meat which has undergone heat treatment or long maturation periods and will therefore generally be tender. On the other hand, inadequate processing methods (too intensive cooking, curing, comminuting) may cause losses in the desired consistency and juiciness, and the best way to check this is by chewing.

## Smell and Taste (Aroma and Flavour)

These characteristics are related to each other to a certain extent because they have to be evaluated together for the reliable determination of a product's flavour. The smell of fresh meat should be slightly acidic, increasing in relation to the duration of the ripening period because of the formation of acids such as lactic acid. On the other hand, meat in decomposition generates an increasingly unpleasant odour owing to substances originating from the bacterial degradation of the meat proteins, such as sulphur compounds, mercaptane, etc.

The freshness of meat is generally indicated by its smell together with its appearance and colour. Sorting out deteriorated meat is mandatory from the point of view of the product's palatability. It is also important because of the fact that high bacterial contamination of meat in decomposition could be accompanied by food-poisoning bacteria(pathogens), which have a deleterious impact on consumers' health. On the other hand, the best fresh meat can also be heavily contaminated with food-poisoning bacteria because these micro-organisms do not cause organoleptic alterations by destruction of meat proteins. Food poisoning can therefore only be avoided by proper hygienic meat handling. The flavour of fresh meat can also be checked by putting small samples (approx. 10 pieces of 1 $cm^3$ each) in preheated water of 80°C for about five minutes (boiling test). The odour of the cooking broth and the taste of the warm meat samples will indicate whether the meat was fresh or in deterioration or subject to undesired influences, for instance rancidity of the meat fat, any a typical meat flavour due to the feed and the sex (boar taint) of the animal or treatment with veterinary drugs shortly before slaughter.

When processing the meat, the smell and taste of the meat products can differ a great deal owing to heat treatment and the use of salt, spices and food additives. Every meat product has its typical smell and taste, and the test person should know about it. Changes in these qualities indicate the use of improper raw materials or a deterioration of the meat product during storage.

Experience is required to become acquainted with the typical flavour (smell and taste) of foods. Only four basic taste components—sweet, sour, bitter and salty—will be perceived by the taste buds. These receptors are small papillae located in certain areas of the tongue. However, the overall flavour consists of smell and taste produced by the meat components and influenced and covered by spices and those compounds produced by ripening or heat treatment. Flavour test panelists should be aware of these special cases. Panelists should not smoke or eat spicy meals before starting the test and should rinse their mouth frequently with warm water during the test.

Sensory evaluation plays an important role in the examination of meat and meat products. Not only does scientific sensory evaluation with skilled panelists using special test programmes and point systems give reliable results, but useful results can also be obtained in a simple way at the consumer level. For the average consumer sensory evaluation is the only way to decide whether or not he or she should buy or eat a certain product.

In developing countries consumers do not receive sufficient information and training on this point, although it is often the only means available for quality control. Sensory evaluation is easy to understand and to perform. What is needed is a basic knowledge of the composition of foods and their typical texture, colour and flavour.

## Physical Test Methods

Physical test methods focus either on the actual condition of meat and meat products, or on the conditions around the product, for example in storage rooms, packages, etc. Equipment will be needed for all these tests which is easily applicable and resistant to utilization under the conditions of practical meat handling and processing.

### *Temperature*

Storage of meat and meat products requires low temperatures to make sure that the growth of micro-organisms will be retarded (chilling between-1 to +4°C) or inhibited (freezing preferably in the range of -18 to -30°C).

Cooking of meat requires high temperatures (starting from a temperature of about 55°C needed for denaturation, but generally higher temperatures are applied, up to 100°C). Canning of meat requires temperatures above 100°C, and for sterilized products where all micro-organisms are inactivated, at least 121°C. These examples demonstrate the importance of different temperatures for different purposes and the necessity of exact temperature measurements using thermometres or temperature recorders.

Glass thermometres should not be used in direct contact with meat because they may break, leaving undesired fragments in the meat, but they are useful when permanently fixed to walls of chillers or production rooms or to cooking equipment or autoclaves for easy checking of the relevant temperatures.

Mechanical bimetal-thermometres, utilizing the extension or contraction of a bimetal spiral under various temperatures, are not very accurate and not sufficiently shock-resistant for practical work in meat industries. Nevertheless, they are widely used and can serve for rough estimates.

Electrical thermometres, consisting of a sensor and a battery-powered electronic instrument, are well suited for meat industries. The sensor functions as a semiconductor. Under different temperatures, differences in the electric conductibility of the sensor are produced. The temperature which the sensor takes up by contact to the surrounding media (water, air, meat, etc.) produces a certain voltage in the electric system. This voltage is registered and displayed as a digital reading of the actual temperature on the instrument.

Advantages of modern electronic thermometres are:

- no glass or other parts that can easily break;
- the sensor can be easily pushed deep into the meat, as well as into frozen meat, and is also heat-resistant under sterilization temperatures;
- display of the correct temperature within seconds;
- no frequent calibration necessary;
- a wide range of temperatures can be covered using one instrument (the temperature range of the instruments recommended for use in meat industries should be between +140°C and -40°C); and
- accurate temperature measurement, also in decimals.

### Humidity

In some special field of meat processing and storage, air humidity is of importance.

In cutting rooms the humidity of the air should be below the level which would cause vapour condensation on the surfaces of the meat being deboned and cut. Vapour condensation may enhance bacterial growth.

Storage chillers for fresh meat require a balanced air humidity that does not cause wet surfaces on the meat with resulting accelerated bacterial growth, but on the other hand keeps evaporation losses low. The relative humidity recommended for this special purpose lies in the range of 70 percent.

Chambers for the maturation of raw hams or dry sausages of the salami type require controlled air humidity, starting from 90–95 percent and after a certain period finalizing the process at 70–75 percent relative humidity. This procedure is important for the balanced drying and ripening of the products. Suitable instruments (hygrometres) for the exact measurement of relative humidity are therefore needed.

In simple but less accurate hygrometres a hair or special synthetic fibre is connected with a pointer and, according to the contraction of the hair or fibre under dry conditions and its extension under wet conditions, the pointer indicates the actual relative humidity on an appropriate scale.

A more accurate way for humidity control is the psychrometric system. These instruments use a dry and a wet sensor to define the ambiental temperature. The temperature indicated by the wet sensor will always be lower, in this case, because of evaporative cooling. The drier the air, the more intensive evaporative cooling will be. Using both temperature values (dry and wet temperature), the value of the relative humidity is determined in practical work using a table for easy calculation.

A modernized psychrometric system which uses electronic devices is available. In this case the humid sensor has not actually to be kept wet, but consists of hygroscopic material with altering electrical resistance. The relative humidity calculated from the temperatures delivered by the sensors by means of a micro-processor is directly displayed on the instrument.

### Water Activity ($a_w$)

Water activity is the free water available for microbial growth in a food product. Free water is that part of the water content that can be eliminated from the product in the form of water vapour. Hence, the term "water activity" is defined as the ratio of the water vapour pressure measured in the product and the pressure of a saturated water vapour atmosphere at the same temperature. This physical definition is used in connection with a number of meat products whose keeping quality depends on their water content. Micro-organisms need a certain degree of moisture to be able to grow on foods. The minimum moisture content necessary for microbial growth varies with the single species of micro-organisms and can be expressed in terms of minimum water activity required.

The lowest $a_w$-values permitting growth of spoilage organisms are:

- normal bacteria 0.91
- normal yeasts 0.88
- normal moulds 0.80
- halophilic (NaCl-tolerant) bacteria 0.77.

The keeping quality of dried meats and meat products without refrigeration depends on their water activity. Dried meat such as biltong, charque, etc. reaches a sufficiently low water activity to be shelf-stable. However, water activity should decrease as fast as possible as slow drying could cause deterioration in a prolonged phase of the process with a still high water activity. The situation is more complicated in the case of products which cannot be dried too intensively such as dry sausages or raw hams. The water activity of these products is relatively low but would still allow the growth of some undesired micro-organisms. Under these circumstances an appropriate shelf-life has to be ensured by the combination of several inhibiting factors, i.e. water activity, content of salt and curing ingredients and the acidity of the product.

Information on the water activity of certain products can be important for further handling, packaging and storage. Simple methods for the determination of water activity are therefore useful.

As water activity refers only to the water available for microbial growth in a product, the chemical analysis of the total moisture content is of limited value since it would also include the water bound by the proteins. The proper way to determine water activity is to measure the humidity of the remaining air in a hermetically closed small cabinet which is to a certain extent filled with the product sample. After a short time a hygroscopic equilibrium between the sample and the surrounding air will be reached. Thus, the humidity determined in the air is equivalent to the humidity available in the product and water activity can be calculated.

For the measurement of air humidity under these conditions, the same principles apply as previously described. Simple devices utilize the extension or contraction of hairs or synthetic fibres, and more sophisticated and more expensive systems use electronic devices.

The sample is placed in the bottom part of the tin and then the lid of the tin that contains the device to measure the humidity is hermetically screwed on. After two hours, hygroscopic equilibrium is reached in the can and the reading of the instrument corresponds to the actual water activity of the product, provided the test has been carried out at a temperature of exactly 25°C. If this temperature cannot be maintained, corrective calculations will be necessary.

Some examples for values of water activity ($a_w$) of different products are shown:

| | |
|---|---|
| fresh raw meat | 0.99–0.98 |
| cooked ham | 0.98–0.96 |
| frankfurter-type sausages | 0.98–0.93 |
| liver sausage | 0.97–0.95 |
| raw cured ham | 0.96–0.80 |
| dry sausage (salami type) | 0.96–0.70 |
| dry meat | 0.75–0.50 |

A certain number of micro-organisms are inhibited at $a_w$ 0.95, but other species are still able to grow. At $a_w$ 0.92 all bacteria groups are inhibited, but the growth of moulds and yeasts is still possible.

### *Airtight Closure of Cans*

For shelf-stable canned meat products two aspects are important from the microbiological standpoint. During sterilization, micro-organisms and their spores have to be inactivated and the can must be hermetically sealed to avoid further contamination of the product after sterilization.

Invisible small perforations of the tinplate or small defects after the closure of the lid will inevitably lead to a *recontamination* by penetrating bacteria and after a short time spoilage of the canned product will occur. Cans should therefore be checked from time to time for these defects.

A simple method is available for this purpose. Using an air-pump with a special device to penetrate the tinplate, the air is pumped into a closed but empty can. The internal pressure built up in the can can be controlled by a manometre connected with the air pump. When dipping the inflated can into water it can easily be seen whether the can is hermetically sealed and if not where the cause for the permeability is, either in the prefabricated body (side wall and bottom) of the can or in the area of the lid seam. In the latter case the function of the can-closing equipment in the processing plant has to be thoroughly checked.

### Weight Differences

The high water content of meat (approx. 70 percent) and meat products (which varies from 70 percent to 10 percent) causes weight differences owing to evaporation losses or drip losses that occur during handling, processing or storage.

Unpackaged meat and meat products are especially subject to considerable evaporation losses. During chilling of warm carcasses evaporation losses of 1 to 2 percent cannot be avoided but further evaporation losses of chilled or frozen meat should not occur when suitable storage conditions (not too dry) or suitable packaging (plastic bags, containers, boxes) are employed. However, some drip losses of packaged meat cannot be avoided.

During meat processing weight losses of meat by cooking, frying or other heat treatment can be registered and reach high values (up to 30 to 35 percent). These losses are unavoidable.

On the other hand, some meat products require weight losses by evaporation to reach their specific keeping quality, for example raw hams, dry sausages or dried meat. In this case, water activity as previously described plays an important role.

Information on weight losses in meat handling and processing is important from the economic and technological point of view. Weight losses can easily be determined using scales of different types, such as suspension scales for carcasses or batches of products and horizontal scales for packages or portions.

### Salt Concentration in Brines

In addition to dry curing methods (dry salt and curing ingredients on the meat), brines are also used for pickling and curing the meat. Brines contain salt and in most cases also sugar and nitrite dissolved in water. With this curing process, meat is either immersed into a brine or the brine is injected into the meat with special devices. In both cases salt is a limiting factor for the sensory quality of the products; in other words salt is needed but should not exceed 2.5 to 3 percent in cooked cured products and 4.5 to 5 percent in raw cured and dried products.

To comply with these requirements, simple methods for testing salt concentration in brines are necessary. For this purpose salimetres have proved to be a useful piece of equipment. Salimetres are densimetres, the graduation showing salt concentrations. Salimetres are dipped into the brine and according to a lower or higher salt content they sink deeper or less deep into the brine. The reading of the salimetres at surface level indicates the salt concentration of the brine. The various technologies of meat curing use brines with NaCI-concentrations in the range of 8 to 22 percent.

### *Other Physical Test Methods*

The physical test methods which have been described can easily be performed since the use of the technical equipment necessary is not too complicated. Other physical test methods exist too, for example, light intensity measurement, colour measurement or texture and consistency measurements of meat and meat products. These tests require rather complicated and expensive instruments and skilled technical personnel. For routine work, criteria such as light, colour, texture and consistency can be evaluated in a satisfactory way by using the corresponding sensory test methods.

### *Chemical Analysis*

Chemical characteristics of foods are related to the product itself and refer primarily to the content of specific substances, which are important from the point of view of keeping quality, flavour, nutritional value, etc., or which may also represent harmful residues.

The test methods necessary are generally complicated and need sophisticated equipment. However, there are also some simple and quick methods for chemical testing with sufficient accuracy which can be applied in the daily routine work such as pH-measurement, moisture/fat/protein determination and various screening methods utilizing test paper strips.

### *pH-measurement*

The pH-value or acidity of meat is important in relation to the meat's microbiological and keeping quality. In the live animal the pH-value of the muscular tissue is about 7.0 to 7.1. Very soon after slaughter a drop in the pH-value is observed and after several hours (24 hours or less) the pH-value reaches its lowest level of about 5.6 to 5.8. The increasing acidity is because of the post-mortem formation of lactic acid from glycogen, a sugar-like substance stored in the live animal's muscles for energy supply. In meat lactic acid causes a decrease in pH-value which is favourable for keeping quality (low pH inhibits bacterial growth) and for flavour (acidity is one of the components of meat flavour).

However, the pH of meat is not uniform either in different carcasses or in different muscles of one carcass. Physiological oscillations do not greatly harm meat quality but abnormal reactions in meat are of great economic, hygienic and technological impact.

There are two types of abnormal reaction with regard to the pH in meat. First the pH-value may drop too fast and second it may not reach the normal low level several hours after slaughter, but remain in the range of 7.

Both abnormalities can easily be detected by pH-measurement in the meat. A too fast pH-value decrease is evident, when one hour after slaughter low pH-values in the range of 5.6 to 5.8 are already reached. This phenomenon occurs only in pigs and the meat remains pale, soft and exudative (PSE). Because of its paleness and wetness (low water-holding capacity), this meat should *not* be used for ham and sausage manufacture (gives dry, tasteless products).

An insufficient decrease of the pH-value, which occurs both in pork and beef, is of hygienic significance because of the lack of building up a certain degree of acidity and suppressing microbiological growth.

This meat also remains close to pH-value 7 after several hours, and is dark, firm and dry (DFD). It should not be used for meat and meat products which have to be stored over a longer period, such as vacuum-packed meat cuts, dry sausages of the salami-type or cured raw hams. However, it is well suited for cooked meat products because of its extremely good water-holding capacity.

It can be seen from this that the pH-measurement is of particular importance for the selection of the raw material for meat processing purposes. Hence, portable electric pH-metres are widely distributed and utilized in the meat industry.

The pH is measured on meat surfaces or in the meat itself, in the latter case by pushing the sensor into the muscle or by means of an incision using a knife.

The sensor consists of a glass electrode filled with an electrolyte (solution of KC1) and a sensitive glass membrane attached at the top. Through the membrane the difference in the hydrogen-ion concentration, which corresponds to the acidity of the meat, is detected and digitally displayed on the attached instrument.

pH-measurement on meat can easily be performed but the following points must be considered:

- the electrode sensor must be completely filled with the electrolyte;
- the instrument must be adjusted daily (calibrated) using two buffer solutions with pH-values 4 and 7;
- after each measurement the electrode must be cleaned using distilled water;
- before each measurement the temperature of the meat, meat product, etc. must be checked and the instrument adjusted accordingly.

## Moisture/fat/protein Determination

Information about the moisture, fat and protein content is essential for the evaluation of the quality of different meats and meat products. Determination methods have changed a great deal in this field in recent years. Revolutionary techniques were introduced using X-rays, infra-red radiation or microwaves in automatic equipment for quick analyses of moisture, fat and protein. These modern methods are time-saving, the results are delivered within minutes or seconds and high numbers of samples can be tested.

However, the equipment is expensive and therefore not suitable for small industries. For routine controls, where not necessarily highly accurate but reliable results on moisture, fat, protein and anorganic components (ash) are needed, cheaper and less complicated methods can be applied. A specially designed laboratory scale, together with some other devices, is required. After homogenizing and weighing the sample, it is fast dried using an infrared beam (or a microwave oven if available).

The weight difference is equivalent to the product's water (moisture) content. The fat is then dissolved using a fat-extracting liquid and removed together with the liquid. The solvent is evaporated. The weight of the residue represents the fat content of the sample. Finally, the sample is charred in a muffle furnace and the weight of the residue is the ash content. Since the sum of the percentages of moisture, fat, ash and protein must be 100, and since the percentage of moisture, fat and ash is known, the protein content in percent is calculated as follows: 100% minus the percent of moisture, fat and ash. This method is not precise, but it is fast, provides useful results about the composition of meat and meat products and can be applied without high costs.

For chemical evaluations a number of screening methods are also available using different test papers. The results are indicated by changes of the colour of certain areas on the paper strips. These test papers are used for pH-measurement, screening of the nitrite content and even for the screening of some harmful residues such as antibiotics. pH-measurements on meat with test strips are negatively influenced by the meat pigment making the colour determination often difficult and the pH-determination not very accurate.

### *Microbiological Examination*

These control methods cannot be carried out without laboratory equipment, because they require sample preparation under sterile conditions, incubation of the samples under constant temperatures and sufficient microbiological knowledge on the part of the personnel involved to interpret the results. However, the application of microbiological methods is the only way to obtain information about the hygienic status of places, equipment and foods. It is true that unclean conditions will always indicate high microbiological contamination and one could argue that a thorough cleaning-up rather than a further microbiological analysis would be needed in those cases. But there could also be the need of detecting the source of permanent contamination (for example through the water, movement of personnel, raw material delivered, etc.) or of food poisoning bacteria. Under these circumstances microbiological examinations can often be very helpful and solve immediate problems.

Some methods suitable for routine work should be mentioned.

### *Trigger Methods*

Microbiological culture media in special small moulds are lightly pressed against walls, equipment (knives, machines), meat surfaces or hands of personnel. The micro-organisms adherent to these objects are absorbed by the surface of the culture media, and after adequate incubation (one to two days at 30 to 37°C), microbial colonies can be identified and counted on the media. Each one of the colonies grown during incubation corresponds to one micro-organism which was on the object tested.

Instead of culture media a special sterile strip of cellotape together with a trigger can be used for taking samples from surfaces. After that the cellotape is laid on a culture media for incubation. This procedure allows the utilization of one culture medium for the incubation of different samples at the same time. However, there is one disadvantage with the trigger system. In the case of high bacterial contamination of the surfaces, tested bacterial colonies will grow very densely together and can no longer be counted.

### *Swab Method*

Surface contamination related to a certain area can be sampled using a sterile swab. After rubbing the swab gently along the surface to be tested, it is suspended on the surface of a culture media. In contrast with the trigger method, bacterial contamination can be spread over the whole surface which is important in the case of high contamination. Thus the samples can always be evaluated since the single colonies are not grown together. However, the method lacks some accuracy since bacteria may remain in the swab.

## Meat Production and Quality

### *Meat and Meat Products*

***Livestock Products as Food:*** Current trends indicate that by the end of the century 80% of the world's population will be living in the under-developed countries and a significant number of these will have large food deficits. An increased production of animal protein would make an important contribution towards filling this deficit (FAO 1984, FAO 1985, FAO 1990A).

On a world-wide basis cereals supply more than 50% of human requirements for energy and nearly 50% of the protein. Animal products, meat, milk, eggs and animal fats, supply 17% of the energy and 32% of the protein but there are vast regional differences between developed and under-developed countries. In Oceania and North America, for example, the amounts of protein available per caput per day from meat are 31.5 and 38.3 g respectively, compared with 4.5 g in the developing countries of Africa and 4.8 g in the Far East (FAO Food Balance Sheets, 1990).

The amounts of protein and fat from meat available per caput per day are shown in table. The ranges indicate considerable differences even between countries classed together as developed or developing.

In most communities meat has long occupied a special place in the diet, for a variety of reasons including taste preference, prestige, tradition and availability, with the nutritional aspects being included more recently (Rogowski 1980).

While it is true that meat is not essential in the diet and many people thrive on diets derived largely or even entirely from plant foods (so long as the amounts and variety are sufficient) there are many diets that would be considerably improved by the inclusion of even small amounts of meat and meat products. This is because, compared with plant foods, they are concentrated sources of protein and a range of vitamins and mineral salts.

As little as 25 g of meat will supply 45% of a child's daily need for protein and half the vitamin B12; the addition of 100 g of meat to the average Zambian diet would increase the protein by 50%, iron by 12%, niacin by 40% and energy by 25% (Jensen 1981). Apart from supplying additional protein the amino acids in that protein complement the cereal sources of protein by making good their relative deficiency of lysine. Moreover, compared with plant foods the iron in meat is well absorbed and meat promotes the absorption of iron from other foods.

## Types of Meat

Meat is the flesh and organs of animals and fowls. There are various legal definitions of meat in different countries designed to control the composition of products made with meat. The flesh of cattle, pigs and sheep is distinguished from that of poultry by the term red meat, while the flesh of poultry (chicken, turkey, duck, pigeon, guinea fowl) is termed white meat.

In addition to the common domestic animals a wide variety of wild animals are eaten - possum, deer, rabbit, moose, caribou, bear, polar bear, seals, walruses depending on availability and local custom, as well as horse, camel, buffalo, goat, dog and rodents. Meat from non-domesticated animals is sometimes termed game meat (de Janvry and Sadoulet 1986). Overall, as indicated, by far the greater part of meat supplies is from four sources but this may not be so in certain localities.

The relative importance of these various sources of meat in the diet varies from region to region and in different cultures; many that are rejected for various reasons in one culture are fully accepted in others. In the Indian sub-continent beef is socially and economically perceived as being second class compared with lamb, mutton and poultry, while the reverse is true in most industrialized countries; many western people abhor the thought of eating dog or horse meat which is relished elsewhere; the relative demand for organ meats compared with muscle meat varies in different regions.

## Production and Consumption

***Developing and Developed Countries:*** Agricultural progress in most developing countries has mainly involved an increase in the production of staple crops and the introduction of industrial crops. New varieties, improved farming techniques, greater use of fertilisers, irrigation and chemical control of pests (a group of procedures collectively termed The Green Revolution) have resulted in considerable increases in production, sufficient, in the absence of climatic disasters, to meet domestic needs in many countries and even, in some instances, to provide a surplus for export.

On the other hand developments in livestock production have lagged far behind. Although there has been an increase over the years in the amount of meat available in developing countries the quantities are small. Between 1980 and 1990 world production of meat increased by 29%; 15.6% in developed countries and 56% in developing countries, but the latter was from a very low base. The daily per caput availability of protein from meat increased by 24% but this was an increase from only 4.9 g to 6.1 g in contrast these figures increased in developed countries by 8%, from 27.4 to 33.9 g per day (FAO Food Balance Sheets).

The mean annual output of milk and meat from cattle in the developing countries of Africa and the Far East is less than one tenth of that obtained in Europe (Blaxter 1975; FAO 1990A). This is a result of traditional subsistence farming practices which provide minimum feeding and management levels to livestock. Animals in developing countries are prized mainly for their draught power and manure world-wide some 250-300 million buffalo make a major contribution to the supply of energy; so far as food is concerned they are prized more for their milk than for their meat. Yet as per capita income rises in Third World countries the demand for meat products is rising faster than that for cereals and outpaces supplies.

The problems of meat production are complex and include multiple biological, economic and social factors. The practices of the small-holder system of livestock production need to be gradually developed so as to fit local conditions and meet increasing demands. Modern scientific practices developed in industrialized countries are rarely directly transferable to developing countries and, if they are to be transferred usually need to be adapted to local conditions.

### *Promotion of Livestock Production*

Three main reasons have been suggested for devoting scarce financial and technical assistance to livestock production in developing countries:

1. The possibility of moving into activities with higher added value per unit of product marketed and into products with higher income elasticity of demand.
2. The possibility of increasing supplies of grain products and of diversification of marginal lands into the production of feed grain, oilseed crops and fodder crops as a result of the Green Revolution.
3. The perception of livestock as a means of increasing rural incomes and increasing rural on-farm and off-farm employment (APO 1976, Bachman and Paulino 1979, De Boer 1982).

### *Animal Versus Plant Production*

Meat production from grazing animal calculated as energy or protein yield per hectare is very inefficient when compared with plant products. Yields, of course, vary enormously from one region to another and even from one farm to another in the same region reveal the comparative inefficiency of animal production in those terms.

Animals are, moreover, poor converters of energy into foods for human consumption; if cereal grain is fed to livestock it requires on average 7 kcal input for every kcal generated - ranging from 16 for beef production to 3 kcal for broiler chickens. One argument that has been put forward against industrial systems of meat production is the competition between animal feed and food for direct consumption by human beings. However, certain animals like ruminants are valuable as converters of inedible agricultural and industrial by-products such as bagasse, molasses, sugar cane rinds, beet pulp, cotton seed hulls, poultry manure and urea, into products of high nutritional value, and they can graze on marginal land that is otherwise of little use (FAO 1976, APO 1990).

This can be illustrated by the considerable production of milk IA India largely using feed materials unsuitable for human food. It is estimated that in the hands of the smallholder some 60% of feed comes from farm by-products and 40% from natural vegetation (Groenwold and Crossing 1975). There have been several publications on this subject - "The Role of Animals in Meeting World Food Needs" (Rockefeller 1975), "The Contribution of Livestock on Small Farms. (FAO 1976), "The Potential for Livestock in Farm Diversification" (APO 1990) - but developments have been slow in most countries of the Third World.

### *Increasing Demand for Meat Products in Developing Countries*

Increasing populations and increased demand per capita together with moderately rapid to rapid income growth lead not only to an increased demand for staple foods but also for preferred foods including, particularly, meat products.

As a result meat consumption grew in the Third World between 1961-65 and 1973-77 at an average rate of 3.4%, and in the fast growth economies at more than 6% (Sarma and Young 1985). The expected average meat consumption by the year 2000 is 20 kg/head/year with a deficit in production of some 20-25 m tonnes. Meat consumption between countries varies from 4 kg per head per annum in low income groups to 35 kg in high income groups. In general there is a relation between income and the consumption of animal products but this does not always hold.

The growing demand for meat products that accompanies rising income has been paralleled by increasing interest in food quality, safety and nutritional aspects, which all give rise to appropriate legislation.

In some countries where economic growth has been rapid and sustained over a considerable period of time the contribution of the livestock subsector to agricultural gross domestic product has increased substantially. In Korea, for example, it increased from 5.4% of agricultural GDP in 1961 to 15.4% in 1973. In Taiwan (Chang 1981) it rose from 18% in the period 1952-57 to 29.5% in 1977. These figures can be viewed against an overall average increase of 1040% in developing countries and of 60% in the United States.

In most developing countries, however, the low level of meat supplies is due to the low return of resources devoted to animal production, which, in turn, depends on the low purchasing power of the vast majority of the population (FAO 1990A).

## Constraints on Meat Production

Other constraints on meat production include inadequate feeding and poor management of animals, to which must be added animal health problems, lack of skilled manpower, problems arising from land tenure, lack of financial resources, and poor rural infrastructure - roads, power, health care, marketing organisation.

There is a great deal of illegal or poorly supervised slaughter which means that regulations are not being enforced, and conditions are unhygienic. Disease, parasites and poor management of all aspects of meat production impede progress. It is estimated that world-wide per year up to 50 million head of cattle and buffaloes and 100 million sheep and goats die from disease and parasites - mostly in the Third World. Even greater losses of production are estimated to arise from ailing and unthrifty animals and from poor handling before slaughter. In many regions of developing countries animals trek to market, often with inadequate feed and water and under considerable stress which lead to losses of both of weight and quality of meat. In parts of Africa, for example, it is estimated that there is often a 30% loss of weight and 10% mortality on the way to market.

In most areas animals are slaughtered on a simple slab, usually at night or in the early morning when the temperature is lower, and the meat is sold without refrigeration or further processing within a few hours. The general disregard for grades and quality of meat both in buying and selling does not encourage investment to improve meat output. A more discerning market tends to develop with economic growth, and both producers and consumers would benefit by sale and purchase on grade and quality bases.

### *Increased Production and Productivity*

Increases in meat production can be encouraged by stable profitable outlets in connection with improved processing and handling facilities and consequently large-scale investment. Loan institutions are usually not geared to make loans to small farmers because the administrative costs of small loans are relatively high and so some 70-80% of small farmers in most developing countries do not have access to institutional credit.

Shortage of trained personnel is another constraint on livestock production; there is a need for skilled producers, processors, distributors, extension advisers and technicians. Lack of qualified control personnel and veterinarians leads to enormous losses in quality and quantity of livestock products.

Lack of refrigeration and other preservation techniques results in considerable losses and can lead to public health problems. If animal production is to compete with crop production intended for direct consumption by human beings, then the efficiency of meat production and meat processing must be greatly increased (FAO 1990B). This is of particular importance for the supply of meat to growing populations.

### *Increasing Yields*

There are many and continuing developments in the western world that increase productivity of animals ranging from artificial insemination of animals with synchronized oestrus and embryo transfer to recombinant DNA technology intended to improve growth and feed conversion. These are far from practicable in most parts of the developing world and, instead, there would seem to be considerable potential in making better use of indigenous animals and more immediate gains from better handling before and during slaughter and from closer control of processing.

The resources necessary for livestock production include water, feed, land, labour, capital and energy. Efficiency of output can be related to any of these and so what is perceived as efficiency will differ with the measure and local availability of these resources. For example, it is common practice in intensive farming in the west to keep species separate and while this is efficient on managed pasture with a sward consisting of a few species of plants it may not be the most efficient method in developing regions since a mixture of types of animals can make better use of the wide variety of grazing species from grasses and legumes through creeping plants to tree-leaves.

One of the biological constraints on the production of animal products is poor food conversion efficiency by the animals, in particular meat-producing animals. Scientific progress has been made in industrialized countries in selective breeding for strains with high feed conversion efficiency that could be adapted to meet the needs of developing countries (Blaxter 1975, De Boer 1982).

### *Age of Slaughter*

One aspect of yield of meat is the age of slaughter in relation to maximum feed conversion efficiency, which applies to animals reared solely for meat. In the early stages when growth is rapid there is comparatively good return of meat for feed consumed. This declines with increasing maturity and at a later stage the weight gain is largely fat - which may or may not be wanted by the consumer but is, in any case, an inefficient procedure.

Thus efficiency of production of livestock products can be improved by appropriate livestock management to capitalise liveweight gain potential of young animals, adequate assessment of marketing premiums for carcass characteristics, together with genetic selection and the use of multi-purpose animals.

### Multi Purpose Animals

While maximum feed conversion efficiency can be achieved by specialising in the production of one product such as milk, eggs or flesh, it may be more efficient overall, despite a small loss of efficiency of production of any one product, to use animals for more than one purpose such as milk and meat or eggs and flesh (multi-purpose animals). It has been shown that dual-purpose beef-milk animals are more energy efficient for the same mix of final products than specialist systems - 44% compared with 34% (Preston 1977).

The choice between specialist and multi-purpose animals will depend on the socioeconomic aspects of the whole process and, to some extent, on the demands of the consumer. Slaughter of culled animals that have completed their life cycle as milk or egg producers results in tough meat typical of old animals, which may or may not suit the consumer. The extreme example comes from the Third World where draught animals are slaughtered when they cease to be economically useful for traction; however, tough meat is often acceptable locally where the meat is cut into small pieces and thoroughly cooked.

### Fuller Use of Animal Tissues

The profitability of livestock production can be increased by making fuller use of the available animal tissues. This requires special attention to separation of the organ meats and to the preparation of by-products. There is often considerable loss of these food materials from spoilage which can be reduced only by greatly improved hygiene and handling. In large-scale processing units with long distribution chains this invariably demands refrigeration. The small-scale sector can manage to market these products through short distribution chains without refrigeration which in any case is not available in most of these premises.

### Unusual Species

There is potential in currently unexploited indigenous animals as sources of meat. Wild animals supplement domestic meat supplies in many parts of the world and there would appear to be considerable potential in developing these animals as managed meat producers. They are already adapted to local environments and so have advantages over imported stock and they appear to be resistant to many diseases that affect domestic livestock. Developments of this kind have already taken place in many countries illustrating this potential e.g. the farming of red deer in Scotland, hybrid deer in New Zealand (Ainger 1991), bison and water buffalo in other areas. Giraffe, elephant, hippopotamus, antelope, rhinoceros and possum can be added to the list; game reserves could be exploited as managed sources of meat.

These indigenous species are unselected and so there is considerable potential in selective breeding for improved growth rates, adequate production per unit of land and improved carcass composition where this is necessary to satisfy the consumer. One comparison between buffaloes and cattle showed that the former gained more weight than the cattle - 0.7 kg per day compared with 0.5 kg - but the dressing-out weight of the carcass was 47% compared with 50% (FAO 1976). The considerable improvements that have been achieved by breeding plans (originally by traditional but more recently by scientifically-based methods) and, above all, by improved overall management in the well-known domestic species indicate the potential achievement if such methods were applied to indigenous species.

It is not possible to arrive at "average" performance of animals since there are such large variations and so much depends on management but comparisons can be made through the "normal" levels of performance. Indicates "normal" levels of performance of some of the commoner domesticated and less common species (Spedding and Hoxey 1975). These animals, of course, may have differing potentials for improvement through feeding, management and disease control.

### Inter Breeding

Some indigenous species will interbreed with conventional livestock and so provide important genetic reservoirs for maintaining and improving the quality of the stock. Their use in meat production could also offer protection to some threatened or endangered species which might otherwise become extinct. Examples are species such as benteng in Indonesia, yak of Central Asia's high country and mithan of the border regions of India, Burma and Bangladesh.

Use is already being made of domestic bovine hybrids; the madura in Central Asia is a hybrid between benteng and cattle; domestic forms of at least two Asian pig species (the Indonesian wild boar and the Sulawesiwarty pig) are important husbandry animals. Other advantages of using unusual species are that some can subsist without encroaching on the feed requirements of others since they eat different species of plants or different parts of the same species, and feed at different times of the day and at different heights from the ground.

### *Buffalo as a Meat Source*

In many countries - Italy, Egypt, Bulgaria, Australia, Venezuela - Buffalo has been developed as an animal resource to produce meat and has been shown to equal and surpass local cattle in growth, environmental tolerance, health and production of meat and calves. Popular misconception about the toughness of buffalo meat is largely due to its consumption after a life as a draught animal; however, when raised for meat and slaughtered at a young age the steaks are lean and appear to be as attractive as beef. Several trials have shown a preference for buffalo steaks over beef steaks; for example the demand for buffalo steaks in the northern territory of Australia exceeds supply and much of the meat in the Philippines is from buffalo. Another plus point is that buffalo meat is accepted by Hindus who forbid the slaughter of cows.

The buffalo is comfortable in a hot, humid environment, and is more resistant to ticks and other diseases. Raising buffalo not only as a draught animal but also as a meat producer could make a major contribution to meat supplies in many parts of the world. It is already exported from India and Pakistan to Thailand, South-east Asia and the Middle East.

At the other end of the size scale is the rabbit, which has a high reproductive rate and yields a quick profit and is free from most social and religious taboos.

## Integration with Crop Production

A consideration of meat production solely in terms of its conversion ratio from crops is incomplete since the development of livestock calls for closer integration between animal and crop production. Livestock and crop activities complement each other at farm level; the crop sector supplies fodder for the livestock (Hudson 1976, Jobling and Jobling 1983), and the livestock provide draught and traction power for crop production as well as manure, and make use of crop residues.

Supplies of fodder can be increased by cultivating varieties with higher yields, fodder can be produced in periods between major cash crops and crop production itself benefits from the application of manure.

Animals yield more than just meat - wool and hair, hides and pelts from the skins, traction power needed for cartage, herding of the animals, irrigation, and transport of the crops, and they produce waste materials that can be used for fuel and biogas production as well as fertiliser.

### Reduction of Losses

***Slaughter:*** In many developing countries the slaughter of animals is traditionally carried out in unsuitable buildings by untrained staff with little attention to sanitary principles. Preslaughter handling is poor and sometimes leads to spread of infection during transportation and in overcrowded lairages, as well as to loss of weight.

The condition of the animal can deteriorate within a few days between selection for slaughter and actual slaughter: fatigue and lack of food will deplete muscles of glycogen which may result in quality deficiencies of the meat after slaughter. While walking animals to market is apparently the cheapest method of transport the loss of weight and the mortality may make this method expensive.

Traditional methods of processing and marketing also militate against quality. Even in larger towns abattoirs that have been specifically designed to supply meat to the expanding centres of urban population too often suffer from unsatisfactory hygiene. Sanitary regulations, where they do exist, may be disregarded and not enforced.

Some of the traditional slaughter operations, developed when they served a small local population, are carried out in the open, on a slab, hanging under a tree or from a fixture in a walled area. There are no cooling facilities and to counter tropical temperatures slaughter is sometimes carried out at night, when it is cooler and the meat transported and sold before mid-day.

In some areas in developing countries the retailer buys live animals from a livestock market or from farms and carries out the slaughter himself and carts the carcass to his retail outlet.

In smaller slaughterhouses this system has to be tolerated but supervision by veterinary authorities is essential. However, in larger operations slaughter should be carried out by a trained staff of abattoir workers in order to maintain the necessary standard of hygiene.

Care must be taken in slaughter and handling since, for example, improper and insufficient bleeding does not allow the necessary degree of acidity to develop in the meat and shortens shelf-life; improper dehiding of the carcass leads to heavy contamination; improper evisceration through accidental opening of stomach and tripes spreads contamination; contact with unclean materials during transport adds to contamination.

### *Refrigeration*

The commonest method of prolonging the shelf life of meat is by cooling. Meat is first chilled after slaughter. It may be kept chilled if there is only a short period of time for distribution. For longer storage periods meat has to be frozen. The shelf life of all typs of unpackaged meat held at chilling temperature, 0°C to + PC or even better -1°C to +4°C, is only between a few days and one to two weeks - depending on the cut of meat, temperature, bacterial load and relative humidity. The shelf life is much longer at freezing temperatures and depends on circumstances such as whether or not it is packaged, type of packaging temperature, etc. Some indication of safe storage times at different freezing temperatures; under optimal conditions shelf life can be even longer than indicated.

Most processed meat products also need to be kept refrigerated from the time of processing until their consumption.

However, the provision of refrigeration presents one of the biggest problems in many areas of developing countries since supplies of electricity are often inadequate. If meat has to be stored then greater use must be made of other methods of preservation.

## Traditional Methods of Meat Preservation

Traditionally foods have been preserved by salting, drying and smoking, methods that have been improved by modern technology.

The simplest and most commonly used method is drying in the open air under the influence of wind and sun. Under favourable climatic conditions a product of good quality can be obtained but otherwise losses from spoilage, infestation and contamination can be excessive; and meat is susceptable to natural predators.

Artificial drying plants which are used in advanced meat processing are energycapital- and technology-intensive, and require skilled labour. They are not suitable for the needs of small-scale producers in developing countries who manufacture small quantities for short periods throughout the year.

An alternative is the type of fuelled mechanical dryer used in humid tropical regions to dry export commodities such as cocoa, coffee and copra. They use wood or charcoal as fuel so their use is restricted to areas where there are abundant supplies of wood.

In other areas the most suitable solution may be to improve existing sun-dry methods or to introduce solar drying, a method by which sun drying is enhanced. The process is best carried out in an enclosed structure so the product is protected from rain, dust, insects and predators and reduces the risk of deterioration of quality and spoilage both during the drying process and during subsequent storage (FAO 1990c).

There are many traditional dried meat products in various regions around the world pastrami in Turkey, Egypt and Armenia, charque in South America, kilishi in Nigeria and West Africa, qwanta in Ethiopia and East Africa, biltong in South Africa. Local requirements, tastes and facilities will obviously influence decisions as to whether dried meat is acceptable and also the most useful process.

### Meat Quality

There are two major aspects of meat quality, nutritional quality which is objective and "eating" quality as perceived by the consumer - flavour, juiciness, tenderness and colour - which is highly subjective.

There are considerable differences between the preferences of individuals including preferences for different cuts of meat, lean or fatty, muscle or organ meats, methods of cooking, etc.

In the industrialised countries the demand for what is perceived as eating quality and also the demand for particular qualities for a range of products from the meat processing industry dictate the breed, feed and management of the animals with intensive rearing and specially formulated dietary supplements and a tendency to slaughter earlier.

On the other hand the demand in most developing regions of the world is for more animal products of almost any kind. The animals live under variable conditions often of rough grazing and grow more slowly, yielding older animals for slaughter; when animals are primarily used for draught they are very old at the time of slaughter. Old animals yield meat that is less juicy and of a quality that differs considerably from that demanded in the industrialised countries.

### *Post-Mortem Changes*

The post-mortem changes that take place when muscle is converted into meat have a marked effect on the quality of the meat.

After slaughter the glycogen in the muscle is converted into lactic acid causing a fall in pH from an initial value of pH 6.8 - 7.3 to about 5.4 - 5.8 at rigor mortis. If animals are stressed immediately prior to slaughter as when they are roughly handled or fight one another the muscle glycogen is released into the blood stream and, after slaughter, is rapidly broken down to lactic acid while the carcass is still warm. This high level of acidity causes a partial breakdown of muscle structure which results in pale, soft and exudative meat (termed PSE) - a condition mostly occurring in pigs. The meat losses some of its water-binding capacity which is so important in certain types of meat processing.

Long-term stress before slaughter or starvation uses up the glycogen so that less lactic acid is formed after slaughter resulting in an abnormal muscle condition in which it remains dark purplish-red on exposure to air instead of a bright red colour. This is termed dark, firm and dry (DFD) in the case of pigs and "dark cutting" in beef. The condition is rarer in lambs. Such meat and products made with it have a pH above 6.0 and spoil quickly since the low acidity favours rapid bacterial growth.

PSE and DFD meat are perfectly safe to eat but limited in their processing capacity. PSE meat has higher drip and cooking losses due to the reduced water-binding capacity (WBC). As well as the pale colour the meat has less flavour than usual.

DFD meat has normal or increased WBC and so is suitable for scalded/boiled sausages and other cooked products but it has poor meat flavour. While there is no remedy for these defects in the meat, DFD and PSE meats can be blended with normal meat for the preparation of products of good quality.

After slaughter as the glycogen in the tissues is exhausted rigor mortis sets in and the whole carcass become stiff. This is due to the contraction of the muscle fibres when the actin filaments of the muscle fibres slide inwards between the myosin filaments so shortening the myofibrils. If the meat is cooked when the muscles are still in rigor it is extremely tough. This condition is prevented by "aging" or "ripening" after slaughter which is achieved by storing the meat until the muscles gradually recover their extensibility and become more tender through partial enzymatic breakdown of the muscles fibres. At this stage rigor mortis is said to be resolved.

Rigor is completed in cattle after 12-24 hours and is resolved by periods that depend on the temperature:- 10-13 days at 0°C, 4-5 days at 10°C, 30-40 hours at 20°C and 10-11 hours at 30°C The process is twice as fast in pork as beef or lamb: it is faster in young animals and slower in "red muscles. That function slowly and continuously in the living animal. "Aging" also leads to improvement of flavour.

Obviously if meat has to be sold within a few hours of slaughter it is still in pre-rigor or rigor, and the tough meat has to be cooked longer with some loss of nutrients.

If lamb, and to a lesser extent beef, are chilled too rapidly after slaughter the muscles may undergo extreme contraction or "cold shortening" which results in very tough meat when cooked. Pork is almost unaffected in this way. Cold shortening does not take place when the carcass is cooled more slowly - the temperature must not fall below 10°C before the onset of the rigor. To achieve this the carcass is kept at ambient temperature for some hours to accelerate rigor and then rapidly chilled or frozen - a process called "conditioning".

Old animals, especially old draught animals, have a high content of tough connective tissue in the muscle and prolonged cooking at a low temperature is needed to soften the meat by hydrolysis of the connective tissue - a fact not always known to consumers.

So it is clear that in many areas conditions militate against good quality meat long transport of animals and poor lairage facilities prior to slaughter reduce the glycogen in the muscles , poor hygiene, high ambient temperature and lack of refrigeration during and after slaughter lead to heavy contamination and growth of microorganisms and considerable losses from spoilage together with dangers of food poisoning. All this can be aggravated by inadequate care of the meat during transport and in the market.

Obviously there is room for improvement in conditions of meat production even for purely local consumption to reduce losses and improve efficiency but if shipment of meat to distant parts is to be considered then it is essential to adopt the sophisticated techniques and methods of refrigeration that are now expected in national and international trade.

## Control of Hygiene and Safety

The safety of meat calls for control throughout the food chain from the farm of origin, and inspection before and after slaughter, to the handling and storage of meat and the products until the time it is consumed.

The responsibility for the production of safe and wholesome meat is shared by the industry and the controlling authority. This requires a controlling authority that is adequately resourced and has the legal power to enforce the requirements and which should be independent of the management of the establishment where the meat is produced.

The Codex Alimentarius Commission has elaborated (besides meat inpection Codes) the Recommended International Code of Hygiene Practice for Fresh Meat (CAC/RCP 111976) and the Recommended International Code of Hygienic Practice for Poultry Processing (CAC/RCP 14-1976) which describe the minimum requirements of hygiene for meat and poultry production.

The application of these Codes can be an important step towards the targets:

a) that the food will not cause infection or intoxication when properly prepared;
b) does not contain residues (of pesticides, veterinary drugs and heavy metals) in excess of established limits;
c) is free from disease;
d) free from obvious contamination;
e) free from defects generally recognised as objectionable;
f) has been produced under adequate hygienic control;
g) fulfils the expectation of the consumer in regard to composition.

The Codex Alimenarius Commission Guidelines include advice on the construction of abattoirs and the facilities required; control of pests, quality of water for cleaning and disinfection; rules of meat inspection and hygienic practices (including supervision by a veterinary inspector).

The Code of Hygienic Practice for Fresh Meat is currently under revision to include a more systematic approach to sanitation and process control, namely Hazard Analysis Critical Control Points.

The growing demand for meat, both per capita and due to growing populations, will increase pressure on slaughter-houses. There is obviously a vast gap between the state of the art in industrialized communities and traditional method in some of the more remote areas of the Third World. Custom-built abattoirs which allow separation of the various stages of the process to prevent cross- contamination and sophisticated techniques of quality control are far removed from slaughter under conditions where energy for refrigeration and adequate supplies of hot (potable-quality) water for cleaning purposes are not available.

Such desirable facilities might be made available in densely populated areas where a regular throughput justifies capital expenditure but it is obvious that these standards must be regarded as long-term objectives in remote areas where slaughter and meat production follow tradition rather than scientific principles.

Indeed, the guidelines concede that traditional practices may permit departure from some of the requirements laid down when fresh meat is produced for local trade.

The Codex Commission recognised that small or relatively isolated abattoirs did not warrant the full-time presence of a veterinary inspector if a veterinary assistant is available for meat inspection. But it was also recommended that meat hygiene in general should be under the control of a veterinary inspector. As regards the Code of Practice for small manufacturers the Commission suggested that this should be left to the discretion of the authorities in each country.

### Hazard Analysis Critical Control Points (HACCP)

As is emphasized elsewhere in this publication an important priority in meat production is to minimise contamination with enteropathogenic organisms during slaughter, dressing and subsequent handling of meat.

A recent development to achieve and ensure safety in food production in general is a systematic approach based on an assessment of the various risks associated with each step of the process. The object is to identify the relative seriousness of risk - Hazard Analysis Critical Control Points.

Critical Control Points are any procedures or locations where control can be exercised over factors that, if controlled, prevent or minimise hazard.

With regard to meat production the HACCP concept systematically identifies potential hazards in the entire chain from animal production to consumption and ranks them according to severity and likely frequency. This covers facilities, equipment and operation and is intended to augment and refine the various codes of manufacturing practice undertaken by industry.

The procedure is intended to enable management to take preventive measures rather than depend on intensive testing of the end-products.

### *Future Strategies on Livestock Production*

It is obvious if livestock production is to be expanded to meet demand and to export that financial arrangements, the use of on-farm resources, production in-puts, veterinary and health services, marketing facilities and resources for research will all need to be improved and expanded. Programmes will need to be suited to local conditions. So far as ruminants are concerned large-scale expansion depends on the availability of land.

In contrast with the developed world much of the meat production in the developing countries is in the hands of small-holders with herds of 2-5 animals, with low productivity and it is improvement in their productivity that will help to fulfil most of the demand rather than the development of large-scale production. It is the opportunistic strategy of resourcepoor farmers to give the animals better feed only when at good milk-yielding age or to increase their strength as draught animals at certain periods of the year. It is then that they tend to compete more directly for food with human beings. Increasing the productivity of each animal is the most efficient way of producing more meat products - doubling the yield (of milk or meat) per animal requires less feed than doubling the number of animals.

A large number of technical policies must be defined in order to establish appropriate livestock development programmes. These involve subjects such as breeding, pasture development, use of non-conventional animals, veterinary programmes, improved farming systems, transport and economic policies including financial production incentives and marketing structure.

## Role of Meat and Meat Products in Human Nutrition

### *Nutritional Value of Meat*

Meat, and other animal foods such as milk, can make a valuable contribution to the diets in developing countries. It has less nutritional importance in industrialised countries where a wide variety of foods of all kinds is available. Many diets in developing countries are based on cereals or root crops and are relatively bulky, especially where fats are in short supply, and this can limit the total energy intake. This is especially true of infants after weaning and young children.

The importance of meat in the diet is as a concentrated source of protein which is not only of high biological value but its amino acid composition complements that of cereal and other vegetable proteins. It is also a good source of iron and zinc and several B vitamins, and liver is a very rich source of vitamin A.

### *Composition*

The animal carcass consists of muscle, connective tissue, fat and bone and some 75% water in proportions depending on species, breed, size, age, etc. The muscle (lean meat) is relatively constant in composition in a given species. The greatest variable in the carcass is the amount of fat, which can range from 2% in some free-living animals to 1540% in domesticated animals intensively reared.

Since the major source of variation in meat composition is the proportion of lean to fat it is useful to consider the composition of lean and fat separately and then to calculate the nutrient composition of the product from the proportions of the two.

With the variations between individual animals of the same species under the same management, together with different managements and errors arising from differing sampling and analytical techniques, there are many and considerable differences in the composition of meat as reported in the scientific literature; provides reasonable average or typical values.

It will be noted that the lean meat of various species has similar values for macronutrients and inorganic constituents. The same is true of the vitamins with the exception of pork which has very high levels of thiamin. Although lean meat has a high water content, about 75%, it is a good source of protein - 20% on a wet-weight basis compared with 8-12% in cereals.

### Influence of Diet

The limited effect of diet on the nutrient composition of lean meat is illustrated in a trial in which the composition of intensively-reared beef (fed barley and protein supplements with grazing ad lib) was compared with extensively-reared (grazing alone) as two extremes of husbandry practice (Harries et al 1968). Analysis of the same muscles from animals on the two systems showed no significant differences in their content of protein, fat, iron, thiamin, riboflavin and niacin. There were greater differences between animals fed on the same system on different farms than between different feeding systems, showing that management practices had the greater effect.

With the exception of vitamin A stored in the liver, diet has little effect on vitamin content, but it has been shown that the addition of thiamin to the diet of pigs can double or treble the amount of thiamin in various muscles (Pence 1945).

### Influence of Age at Slaughter

As animals grow the proportions of total nitrogen and fat, and also the amounts of iron, increase as the animals approach maturity, and more slowly after that. At the same time the ratio of polyunsaturated to saturated fatty acids falls. Collagen (connective tissue) becomes less soluble and less digestible so animals that are poorly fed and so take several years to reach a useful size, provide meat of lower eating quality. Animals killed after a lifetime of work provide even tougher meat. Older animals have a high proportion of water-soluble extractives in the muscle and animals reared on poor pasture, and which are therefore relatively old by the time they reach a size suitable for slaughter, have long been used for the preparation of meat extract.

### Protein

The protein of typical mammalian muscle after rigor mortis but before post-mortem degradative changes contains about 19% protein: 11.5% is structural protein - actin and myosin (myofibrillar), 5.5% soluble sarcoplasmic protein in the muscle juice, 2% connective tissue (collagen and elastin) encasing the structural protein and about 2.5% fat dispersed among the protein fibres. Myoglobin is present in relatively large quantities in heart muscle because of heavy oxygen demand: (the highest amount of myoglobin in mammals is found in the whale to permit prolonged submersion under water).

Collagen differs from most other proteins in containing the amino acids, hydroxylysine and hydroxyproline and no cysteine or tryptophan. Elastin, also present in connective tissue, has less hydroxylysine and hydroxyproline. Hence cuts of meat that are richer in connective tissue have lower protein quality. Their content of connective tissue makes them tough and in many regions these cuts of meat bring a lower price.

Immediately after rigor mortis there is almost 2.5% carbohydrate present - lactic acid, glucose and derivatives. As distinct from the average of the whole animal carcass the composition of meat as cut for consumption shows some variation depending on the cut. The types of cut often differ in different regions.

In addition there will be further differences in the amount of water and fat lost in different methods of cooking.

### Lipids

Lipids (fats) are found at three sites in the body.

1) The largest amount by far is in the storage deposits under the skin and around the organs. This constitutes the obvious, visible fat in a piece of meat and can be as much as 4050% of the total weight in fatty meat or fatty bacon. This adipose tissue is composed largely of triglycerides contained in proteinaceous cells with relative little water.

   Clearly this visible fat can be trimmed off the meat during processing, before cooking or at the table - a growing practice in the western world.

2) Small streaks of fat are visible between the bundles of muscle fibres, intermuscular fat, i.e. in the lean part of the meat; this is known as "marbling" and can amount to 4-8% of the weight of lean meat.

3) There are small amounts of fat within the muscle structure - intra muscular or structural fats - in amounts varying with the tissue. This can be 1-3% of the wet weight of muscle nd 5-7% of the weight of the liver.

Structural fats are largely phospholipids and include long chain fatty acids. Fatty acids are of three types.

(1) Saturated fatty acids in which all the carbon atoms in the chain carry their full quota of two hydrogen atoms and the carbons are linked by a single bond;

(2) mono-unsaturated fatty acids (MUFA) in which one hydrogen is missing from each of two adjacent carbon atoms which are therefore linked by a double bond; and

(3) polyunsaturated (PUFA) in which two or more pairs of hydrogen atoms are missing and there are several double bonds in the carbon chain.

The physiological significance of these fatty acids in the human diet is discussed. Species, breed, sex, age, and environment influence the amount as well as the degree of unsaturation of the fat (mainly the ratio between unsaturated oleic acid and the saturated palmitic and stearic acids).

Animals living in woodlands in Uganda and Tanzania (eland, hartebeest, giraffe, buffalo, warthog) and also free-range cattle have only about 2% of lipid in the muscle, of which about 30% is PUFA. Those grazing on grassland have about 3% lipids in the muscle of which about 15% is PUFA.

In contrast lean domesticated cattle (fed on supplemented diets) have about 5% lipids of which only 8% is PUFA, and intensively-reared fatstock have 15-30% lipids (Crawford 1975).

The cholesterol content of meat is compared with that of some other foods.

### Vitamins

The body content of most vitamins is largely independent of diet. Apart from the thiamin effect on pig meat mentioned above, the exception is vitamin A which is stored in the liver in amounts depending on intake, with small amounts present in the kidney - these are the only tissues to contain significant amounts of this vitamin (there are traces, 10- 60 ug/100 g, in muscle).

Under free range conditions of grazing there is a very high intake of carotene (pro-vitamin A) which is mostly converted into retinol (vitamin A). Tables 2-2 and 2-9 list the typical vitamin content of raw meat and offals.

Pig meat is very rich in thiamin compared with all other animals, nine times as much, but has the same content of riboflavin as others.

Liver is by far the richest of animal tissues in all the vitamins, and includes unchanged carotene as well as being the only tissue to contain more than a trace of vitamin D.

### Minerals (Tables 2-2,2-3 and 2-9)

Meat and offals contain a wide variety of mineral salts. The contents of iron, zinc and copper vary considerably in different species, liver being by far the richest source of these minerals compared with muscle tissue.

High levels of minerals in the feed do not necessarily increase the level of that mineral in the flesh and there is a complex inter-relation between minerals. For example, the molybdenum content of sheep meat increases with dietary molybdenum only when dietary sulphate levels are low. Dietary molybdenum inhibits the accumulation of copper which is partly off-set by increased manganese. Liver copper decreases and molybdenum increases with increasing amounts of molybdenum. Other inter-relations between minerals include calcium and zinc (Byerly 1975).

Copper is used in some feeding systems for pigs as a growth stimulant and can result in levels of several hundred parts per million of copper in the liver.

When pasture is deficient in minerals, especially phosphorus and cobalt, the amounts in the muscle are reduced.

### Meat by-products

The amount of carcass meat obtained from animals varies with the type of animal only about one third of the total weight of cattle and lambs and half of the pig.

The other parts of the animal - liver, heart, brains pancreas (gut sweetbread) thymus (chest sweetbread), tripe, feet (trotters), tail, testes (fries), intestines (chitterlings), cheek meat and head meat and fat (tallow, lard, suet) - are collectively called offal, variety meats, side meats or organ meats in various countries. With regard to poultry, the term giblets means liver (with gall bladder removed), heart and gizzard and any other material considered as edible by the consuming country. Not all parts are eaten depending on consumer acceptance, religion and tradition as well as regulations imposed for reasons of hygiene.

Intestines are used as containers for sausages of the different types, blood may be used in sausages, pork skins may be eaten or used as a source of gelatin. In addition, some inedible by-products such as bonemeal can be used as a mineral supplement in animal feed and there are other inedible by-products of economic value such as hides and horns.

## Nutritional Value

In general they are richer than lean meat in iron, copper and certain B vitamins, with liver being a particularly rich source of vitamins A, B1, B2, B6, B12, niacin and pantothenate and even some vitamin C.

Kidney is a rich source of B1, B2 and B12: pancreas is a good source of B1, B2, C and pantothenate.

The vitamin C in lungs, spleen and thymus is usually present in sufficient quantity to allow some to survive cooking.

Other organ meats compare well with lean meat as sources of the vitamins, and all meat products are good sources of zinc and of iron, liver, lungs and spleen being especially rich in iron (Anderson 1988).

Ears and feet have a high protein content but much of this is collagen and so of poor nutritional value, although when consumed this has no significant effect on the quality of the protein of the diet as a whole.

### *Cooking*

Meat as purchased may include bone, outer layers of fat, gristle and tendons which are removed to differing extents before cooking, so that the composition of meat "on the plate" can vary enormously.

Meat and meat products are considered cooked when the centre of the product is maintained at a temperature of 65-70°C for 10 minutes since the proteins will then be coagulated and the meat tenderised by partial hydrolysis of the collagen. The vegetative form of bacteria, but not spores, will have been destroyed (thermoresistant spores can survive heating above 100°C). The completion of the cooking process is generally indicated by a change of colour from red to brown (red to pink in cured products) and flavours are developed.

Denaturation of red myoglobin and conversion to brown myohaemochromogen starts at 40°C and is almost complete at 80-85°C (Lawrie 1991). Cooked flavour results from a number of reactions including changes in lipids, carbohydrate and protein, with heat breakdown of peptides and amino acids and reactions between proteins and carbohydrates.

Meat from older animals richer in connective tissue requires longer cooking at 5060°C - a temperature at which collagen can be hydrolysed. If heated for long periods at temperatures above 80°C amino acids begin to decompose with the production of unpleasant flavours. (Hydrolysis of collagen is rapid during the canning process when high temperatures are employed for only a short time).

In comminuted meat products, such as sausages, the particles of meat become bound together during cooking through coagulation of extracted proteins. In products that contain pastry this has to be cooked at the same time as the meat.

Water is lost during cooking, the amount depending on time, temperature, method of cooking, size of sample, heat penetration and composition leading to an increase in concentration of the fat and protein. The changes in composition and indicates changes in fat content which depend on the method of cooking. There is a loss of water-soluble vitamins, minerals and protein in the juices but this is a small proportion of the total present and, moreover, in most cooking procedures the juices are usually consumed with the meat.

With so many factors that can influence changes on cooking literature data are rarely comparable - unless the work has been carried out in the same laboratory - and cannot be expected to do more than indicate the general effects.

The Massachusetts Data Bank has tabulated the average amounts of nutrients in meats of various types and the large coefficients of variation illustrate the impossibility of attempting to provide precise figures.

### Effect on Fat

Even in deep frying there is a loss of fat since lean muscle does not absorb the cooking fat. As an example from one set of food composition tables, raw rump steak containing 18.9% protein and 13.5% fat has a total of 32.4% dry matter (ignoring minerals). Expressed as a proportion of dry matter this is 58.3% protein and 41.6% fat.

When grilled, i.e. cooked by radiant heat with added fat the loss of water and fat reduces the total fat to 30.7% of the dry matter while the protein increases proportionately 693%.

When fried the loss of water is greater than in grilling but the loss of fat is less so that the proportion of protein becomes 66.2% and the fat 33.7% of the dry matter.

The data for chicken and also shows the differences between dark and light meat and the effect of including the skin in the analysis. Boiling of chicken causes a greater loss of water than roasting but no loss of fat so that as a proportion of dry matter, fat is highest in the boiled product.

### *Effect on Protein*

Proteins can be damaged from the nutritional point of view when part of an essential amino acid is rendered unavailable. This involves first lysine at temperatures around 100°C; then cystine and methionine at temperatures around 120°C, and other amino acids after prolonged heating (Bender 1978).

At the rather low temperature needed to cook meat there is little loss of available lysine and no loss of methionine and cystine. For example no change in protein quality was found after roasting in an open pan at 163°C when the internal temperature did not rise above 80°C; or when the meat was browned in an oven for 30 min. then sterilised in a can (Mayfield and Hedrick 1949, Rice 1978).

When meat is roasted the outer part reaches a high temperature and turns brown because of a reaction between the lysine and reducing substances present (Maillard or browning reaction) which produces the desired roast flavour. However, since the roasted part is only a small fraction of the total piece of meat and the internal temperature does not exceed about 80°C there is no measurable change in the quality of the protein as a whole.

### *Effect on Vitamins*

***Thiamin:*** One of the most sensitive vitamins is thiamin; it is both water-soluble and heat-labile. It is also damaged by oxygen and at neutral and alkaline pH. It is very susceptible to destruction by sulphur dioxide and sulphites which are used in some countries to preserve comminuted meat products. There is also some destruction during treatment with ionising irradiation but this can be reduced by irradiating in the frozen state.

The juices exuded from meat during cooking include part of all the water-soluble constituents, including mineral salts, proteins and vitamins but except for the heat-labile thiamin these are recovered in the juices consumed with the meat, unless they have been damaged by excessive heat.

In a general review of the subject (Karmas and Harris 1988) losses of thiamin are given as 1540% on boiling, 40-50% on frying, 30-60% on roasting and 50-70% on canning. All figures listed for cooking losses of vitamins must be regarded as rough average values since they will depend on time and temperature and conditions of cooking, the particular product, the size of the pieces of meat and thus heat penetration, to which must be added errors due to sampling and the considerable errors that are unavoidable in vitamin determination. Literature figures can be used only for guidance and if more accurate figures are required they must be determined on the product in question subjected to the specific process - and even then are subject to the problems of analysis.

### Riboflavin and Niacin

Cooking losses of riboflavin average around 10%. Riboflavin is relatively stable to most cooking practices (excluding the high temperature of roasting) and to canning and dehydration. It is damaged by sun-drying and under any alkaline conditions; dry-curing and smoking lead to about 40% loss, wet curing to about 10% loss.

Niacin is stable to heat, light, oxygen, acids and alkalies and also to irradiation but can, of course, be leached from the food; losses average about 10%.

### Other B Vitamins

There is less reliable information about other B vitamins but some reported figures are presented. On average about one third of the vitamin B6 and pantothenate are lost in cooking.

### Methods of Processing and Preservation of Meat

Meat was originally processed to preserve it, but since the various procedures cause so many changes in texture and flavour it is also a means of adding variety to the diet. Processing also provides scope to mix the less desirable parts of the carcass with lean meat and in addition is a means of extending meat supplies by including other foodstuffs such as cereal in the product.

Meat is a highly perishable products and soon becomes unfit to eat and possibly dangerous to health through microbial growth, chemical change and breakdown by endogenous enzymes.

These processes can be curtailed by reducing the temperature sufficiently to slow down or inhibit the growth of micro-organisms, by heating to destroy organisms and enzymes (cooking, canning), or by removal of water by drying or osmotic control (binding the water with salt or other substances so that it becomes unavailable to the organisms). It is also possible to use chemicals to inhibit growth and, very recently, ionising radiation (however, the last is not allowed in some countries).

Traditional methods that have been used for thousands of years involve drying in wind and sun, salting and smoking. Canning dates from early in the 19th century and allows food to be stored for many years since it is sterilised and protected from recontamination.

### Chilling and Freezing

While mechanical refrigeration is a modern process it is known that the ancient Romans kept food cool with ice. "Chilled" meat is usually stored at temperatures around 1°C to +4°C when it keeps well for several days. Provided that the meat is kept very cool(1°C to 0°C) and that slaughter and meat cutting are carried out under strict hygienic conditions, modern packaging techniques including storage under carbon dioxide or nitrogen or in vacuum can extend this period to about 10 weeks.

Chilling at temperatures very close to the freezing point of meat, -15°C, diminishes the dangers of most pathogens and slows the growth of spoilage organisms; growth of some organisms, moulds, virtually ceases at -10°C.

Most pathogens (Salmonella, Staphylococcus species and Clostridium perfringens) are inhibited by cooling but Listeria monocytogenes can grow at + 2°C, some Salmonella species at +5°C and Campylobaner at +7°C.

Non-pathogens include Pseudomonas species which predominate on the exposed surface of chilled meat and Laaobacilli on vacuum-packed meat.

Freezing - commercially at -29°C and domestically at -18°C - is now a standard method of preserving for periods of 1-2 years but there is some deterioration of eating quality compared with fresh or chilled meat.

However, there are problems in chilling and freezing meat. If it is cooled too rapidly below 10°C before the pH of the muscle has fallen below a value of about 6, the muscle fibres contract (cold shortening) and the meat is tough when cooked. This problem applies more to small animals, such as lamb, which cool down rapidly. The modern procedure is to cool the carcass to 10-15°C ("conditioning") and to hold that temperature for a few hours until the pH has fallen to 6. Beef carcasses can be suspended in such a way as to exert a pull on certain muscles to prevent contraction. Another method is to apply electrical stimulation to the carcass after slaughter (low volt) or after evisceration (high volt) for 2-4 minutes to bring down the pH rapidly.

Another problem can arise during thawing of pre-rigor frozen meat when the muscle contracts and exudes a substantial part of its weight as tissue fluids (thaw rigor) (Lawrie 1991). Clearly, freezing of meat is not a straightforward procedure and calls for certain expertise. Only post-rigor meat should be frozen.

## Nutritional Changes by Freezing

Meat is frozen without any prior treatment, unlike vegetables which have to undergo a preliminary blanching process to destroy enzymes involving considerable loss of water-soluble nutrients. So there is little or no loss of nutrients during the freezing procedure, nor, so far as there is reliable evidence, during frozen storage - apart from vitamin E.

Proteins are unchanged during frozen storage but fats are susceptible to rancidity. Pork and poultry meat are more susceptible since they are richer in unsaturated fatty acids than other meats, and comminuted meat is also very susceptible to rancidity because of the large surface area which is accessible to oxygen.

The vitamin E is damaged because the first products of fat rancidity, hydroperoxides, are stable at the low temperature and oxidise the vitamin. At room temperature they break down to harmless peroxides, aldehydes and ketones, so that vitamin E is more stable at room temperature than during frozen storage.

The losses incurred in frozen meat mostly take place when the meat is thawed, and juices are exuded containing soluble proteins, vitamins and minerals. This is termed "dripthaw" and the amount depends on the length of time of ageing (time between slaughter and freezing), whether frozen as carcass or meat cuts, conditions of freezing and speed of thawing; it varies between 1% and 10% of the weight of the meat and is usually about 5%.

There is some loss of nutrients when the meat is cooked after thawing; results published in the scientific literature tend to measure the combined losses from the original fresh meat to the final cooked product. Unfortunately the results vary so much that it is not possible to draw conclusions.

It must be emphasised that the variations are largely due to difficulties in analysis of the B vitamins, and to differences in conditions and methodology - even results from the same laboratories are inconsistent. This is illustrated very clearly by results published from one group of investigators who examined pork loin after freezing and storage at -12°C and 24°C and subsequent cooking at regular intervals over one year for changes in thiamin, riboflavin and pyridoxine (Mikkelsen, Rasmussen and Zinck 1984). Despite constant experimental conditions analyses at two monthly intervals showed wide fluctuations, especially for thiamin, which were attributed by the authors to difficulties in analytical methods.

It was tentatively concluded after storage at -12°C and cooking that about 90% of the thiamin was retained but no firm conclusions could be drawn about other vitamins. No conclusions could be drawn about storage at the lower temperature!

For riboflavin about 90% was retained at -12°C and 100% after storage at -24°C and cooking, although these results were also variable.

For pyridoxine 80% was retained when stored at -12°C and cooked but the results were erratic.

In the same report ground beef was examined only after 1 year storage and showed 80% retention of thiamine, 85% of riboflavin and 100% of pyridoxine at both temperatures.

A summary of earlier work (Fennema 1975) suggested that losses during freezing and storage of meat and poultry for 6 - 12 months at -18°C but excluding subsequent cooking, ranged between zero and 30% for thiamin, riboflavin, niacin and pyridoxine. A survey of frozen meals analysed after freezing, storage and cooking reported losses of up to 85% of thiamin, 55% of vitamin A, 33% vitamin E, 25% niacin and pyridoxine (De Ritter et al 1974).

Little research in this area has been reported in recent years and this limited number of reports illustrates the difficulty of making even generalisations about the stability of vitamins in frozen meat products.

### Processing - General Aspects

Processed meats are products in which the properties of fresh meat have been modified by the use of procedures such as mincing, grinding or chopping, salting and curing, addition of seasonings and other food materials, and, in many instances heat treatment. Most of these processes extend the shelf life of meat. Their manufacture, in most instances, depends on the ability of the mixture to retain water since they are emulsions of protein, fat and water.

## Meat Content of Processed Meats

Where there is a demand for consumer protection it is often necessary to legislate to control the meat content of products that include other food ingredients.

Even if no additives are included meat products can contain variable amounts of lean muscle tissue, fat and connective tissue. A method of assessing the apparent meat content of a raw product is by determination of the total nitrogen content on a fat-free basis and multiplying by an average conversion factor, corrections being applied for the contributions from cereals or other nitrogen-containing ingredients e.g. 3.45 for pork products, 3.55 for beef, 3.7 for chicken (3.9 for breast of chicken and 3.6 for dark meat), 3.45 for ox liver, 3.65 for pig liver (3.55 for liver of unknown origin), 2.7 for kidney, 3.2 for blood and 3.0 for tongue. There are no agreed factors for conversion of nitrogen for cooked, cured or processed meat and the apparent meat content of such products is expressed approximately in terms of "raw meat equivalents" (Egan, Kirk and Sawyer 1981).

Other methods used for legal control purposes refer the composition of meat products directly to the nitrogen or protein content of the dry, fat-free product or to the water: nitrogen ratio. Lean meat can be determined directly by measurement of 3-methyl histidine which is characteristic of meat protein but if large amounts of fillers and binders are present the method is unreliable.

The proximate composition of some processed meat products is given in Tables 216, United Kingdom products, and 2-18, United States products; in both instances the composition is regulated so these data are not universally applicable and serve only as examples.

## Curing

Curing was originally a term applied to preservation in general but is now restricted to preservation with salt (sodium chloride) and sodium or potassium nitrite or nitrate or a mixture of these two salts. The nitrate serves as a reservoir for nitrite - the active compound - since bacteria in the curing solution form it from the nitrate.

The use of salt is one of the oldest methods of preserving meat since at concentrations greater than 4% in the aqueous phase it inhibits the growth of most spoilage organisms. To function as a complete preservative the salt concentration would need to be around 17%, at which levels the product would be unpalatable. In most cured meat products the salt concentration is between 2.5 and 5% and the nitrite inhibits the growth of other organisms. Nitrite also reacts with proteins when heated to form compounds (called Perigotype factors) that inhibit the development of spores of Clostridium botulinum, the cause of botulism, the most serious type of food poisoning.

Additionally, nitrite is broken down to nitric oxide, which reacts with the red colouring matter in muscle, myoglobin to form deep-red nitrosomyoglobin. As the protein is denatured, this is converted, rapidly when heated and more slowly otherwise, into a pink compound which is responsible for the typical colour of cooked ham, canned luncheon meat, franfurters and raw ham, dry sausage, etc.

The early curing procedures were lengthy and recent developments have led to a reduction in the time required. For example instead of simply immersing the meat in brine it is first injected with the curing solution and the process can be completed in 1 - 2 weeks. Thin slices of meat such as bacon can be cured in a few hours, and the processing time can be reduced to a few minutes if heated and if the cure is completed in the final package.

Animal experiments have shown potential risk from nitrosamines formed from nitrite but, as discussed, the amounts present do not appear to be harmful to human beings.

The addition of sodium ascorbate to the pickling brine accelerates the curing process because of its reducing capacity and allows smaller amounts of nitrite to be used, so there is less residual nitrite in the meat which reduces the possibilities of the formation of nitrosamines. Residual ascorbic acid has an antioxidant effect in stabilising the colour and preventing rancidity.

### Tumbling and Massaging

A new technique was developed in the 1960's to accelerate the penetration of salt. Pieces of meat are injected with the curing salt solution or chopped meat immersed in it and then mechanically shaken - "tumbled". Solutions of 2-8% salt are used, sometimes with the addition of polyphosphate, when there is some extraction of water-soluble protein, mainly myosin. The effect is to improve the water-holding capacity of the meat by reaction between the salt and the structural proteins, aided by the polyphosphate. The extracted proteins set to a strong gel on heating and so bind together the pieces of meat, which can then be shaped or sliced.

The term "massaging" is applied to a relatively gentle mechanical treatment while "tumbling" is a more vigorous action.

## Smoking

Meat has been treated with smoke from the earliest days - traditionally over a wood fire and more recently by producing smoke from wood sawdust in a generator and conducting the smoke over the meat.

The substances deposited on the meat contribute to the flavour and appearance but with ordinary, light smoking the preservative effect is limited and the product has to be stored refrigerated. Intensive smoking does prolong shelf life both by heavier deposition of preservatives and by the drying effect of the hot air but it has a detrimental effect on flavour. Consequently preservation by smoking is regarded as an emergency measure when other methods cannot be used.

A modern development making use of the flavouring effect is to use an aqueous solution of the constituents of smoke which reduces the amount of strongly flavoured and other unwanted substances.

## Processed Meat Products

***Common Cured Meat Products:*** The commonest cured products are sausages, bacon, pork shoulder, ham, luncheon meat; any type of meat can be cured either as whole cuts or after comminution.

Bacon is cured pork, in various countries traditionally made from specified parts of the pig but it can be made from any part. There are modifications of the process including so called sweet cure with added sugar (0.25%) and mild cure with less salt.

Bacon can vary greatly in the amount of fat and there are considerable differences between the various published figures; those shown are from the same source and so are comparable with one another.

Ham is the cured product of the upper leg and buttock of the pig and differs from gammon only in that the latter is cut from the side of bacon after it has been cured. It is stable when raw after a certain period of maturation but is often cooked to pasteurisation temperature, 70°C, or it may be canned at pasteurising temperature. It may be smoked as an additional means of preservation and flavouring.

Typical analysis of canned ham per 100 g: 65-72 g water, 18 g protein, 5-12 g fat, 0.5-0.8 MJ, 1100-1250 mg sodium, 1.2-2.7 mg iron, 0.2 mg copper, 2 mg zinc, 0.5 mg thiamin, 0.2-0.25 mg riboflavin, 4 mg niacin, 0.2 mg vitamin B6, and may have residual ascorbic acid 10-60 ma.

## Sausages

There are some 800 types of sausage made of comminuted or chopped meat of various kinds, seasoned with salt and spices, often mixed with cereal and packed into natural casings (consisting of the connective and muscle tissue of animal intestines) or made of cellulose, collagen or synthetic materials. There are six main types of sausage - fresh, smoked, cooked, smoked and cooked, semi-dry and dry.

Frankfurters, Bologna, Polish and Berliner sausages are generally made from beef, pork and pork fat comminuted with the addition of curing salts and are smoked and cooked. Thuringer, soft salami, mortadella, and soft cervelat are cooked and semi-dry; pepperoni, chorizos, dry salami and dry cervelat are slowly dried to a hard texture without cooking. The nutrient content of a number of products is given. There are several variants of each type of sausage; for example, a US table of food composition (Watt and Merrill 1975) includes six types of Frankfurters, namely all meat, with non-fat dried milk with cereal and with both these additions, also raw and canned.

Great Britain which differ considerably from the US figures. These differences can be attributed to variations in composition and method of analysis, and serve to illustrate the approximate nature of any tables of nutrient contents of processed meat products. A major reason for the difference in thiamin content in some comminuted meat products between US and UK tables, is the use of a preservative in UK, namely sulphur dioxide, which destroys most of the thiamin.

Liver sausage contains 10-20% liver and in many cases other edible offals. Blood sausage contains 10-20% whole blood with nitrite salt (not precooked). Other components are precooked meat, edible offals, fatty tissue (cooked sufficiently to separate fat with a low melting point) and pigskin. This type of sausage has a firm consistency due to swollen connective tissue and gelatinized collagen.

Fermented sausages are dry sausages including salami, dry pork and beef sausages and summer sausages, that have been subjected to bacterial fermentation. Meat from a variety of animals may be used, including camel, donkey and horse but rarely mutton, goat or venison. Only well-chilled or frozen meat is used and a temperature of -2 to +5°C maintained during chopping to facilitate comminuting of lean and fatty tissues to the particle size desired and to avoid deposition of fat drops in the batter. Added salt prevents the growth of unwanted micro-organisms and extracts salt-soluble proteins to form a protein gel which binds the pieces of meat together. The bacteria originate from the natural flora of the meat and the environment although starter cultures of Micrococcus, Pediococcus cerevisiae, etc., are sometimes used.

During the slow, prolonged fermentation the pH falls to between 4.8 and 5.4 then the product is dried and may be smoked. Fermented sausage is not cooked and preservation depends on the high acidity and high salt content toegther with the low water content.

### Additives

Comminuted products such as sausages and luncheon meats are based on lean meat, which, technologically, provides water-holding and meat-binding capacity, with the addition of fatty meats and, sometimes, organ meats. The amount of these is limited otherwise the products have an unattractive soft texture and high shrinkage on cooking.

The ingredients include cereals and potato starch, termed fillers, which also serve to bulk out the supply of meat products ("meat extenders"). Other ingredients include a number of substances which have considerable water-holding and binding capacity ("binders"). These include egg or egg yolk blood plasma, skim milk powder, caseinates, soya isolates, wheat gluten, whey protein and dehydrated products derived from various vegetable proteins (soybean, safflower, corn, peanut and pea protein) and their binding properties depend on their ability to form irreversible gels on mild heating which serve to hold together the small pieces of meat.

It is not posssible to generalise about the nutritional value of products of such variable composition.

### Other Comminuted Products

These are made from chopped (minced, ground) meat and fat and usually include meat from various parts of the animal including trimmings. In specific products organ meats are also used. Some tissues can be included only in restricted amounts because their texture can adversely affect the product.

Other ingredients are included for the purposes of ekeing out meat supplies or for their capacity to bind the minced pieces and seasoning.

Seasoning is a comprehensive term for ingredients intended to improve flavour such as salt, pepper, spices, herbs and vegetables. Spices commonly used include cinnamon, cassia, clove, ginger, mace, nutmeg, paprika, cardamom, coriander and mustard; herbs include sage, savory, bay leaves, thyme and rosemary; onions and garlic are also used.

Liver and blood have a pronounced colour and flavour and can be included in comminuted meat products only in limited amounts (15-50% of the total) depending on the local acceptability or may be used in specific liver or blood sausages.

Pates can be included with comminuted products since they are made from coarsely or finely chopped meats, precooked and seasoned, and some are cured with salt, nitrite and phosphate.

### Luncheon Meat

This is the name given to several products made from finely chopped meat and fat with the addition of cereal and water, preserved by the addition of salt and nitrite and by heating.

Although the nutrient content is very variable a typical analysis of a canned product of this type, per 100 g is:- 5 g water, 13 g protein, 27 g fat (10 g saturated, 12g monounsturated, 3 g polyunsaturated, 70 g cholesterol), 1.3 MJ, 1000 mg sodium, 1 mg iron, 0.3 mg copper, 2 mg zinc, 0.07 mg B1, 1.8 mg nicotinic acid, 1 mg B12.

### Corned Beef

This was originally a by-product from the manufacture of meat extract when it was the only meat product that could be shipped from South America before the introduction of refrigerated transport. The latter is made by hot water extraction of the low quality meat from animals that have taken several years on relatively poor pasture to reach suitable size so is relatively rich in connective tissue (as well as in muscle extractives).

The meat is coarsely chopped and immersed in hot water to extract the solubles; the exhausted meat is cured by the addition of coarse grains of salt ("corns") and nitrite and canned, often with the addition of fat. It has an extremely long shelf- life and used to play a major role in military rations and in expeditions.

Other meats such as mutton can be treated similarly and some modern processes use unextracted beef.

In the United States the term "corned beef" is applied to what is elsewhere termed "salt beef" i.e. cured whole beef.

Since corned beef is made from extracted meat it is low in water-soluble vitamins, containing per 100 g only a trace of thiamin, 2.5 mg nicotinic acid, 0.2 mg riboflavin, 2 mg B12, 3 mg iron, 0.25 mg copper, 6 mg zinc, together with 950 mg sodium and 60 g water, 26 g protein, 12 g fat and supplying 0.9 Ml (220 kcal).

### Burgers

Patties made from minced meat have become popular world-wide through the agency of international fast-food outlets. These were originally hamburgers made from beef with the name derived from the Hamburger sausage but may be made from any meat and "muttonburgers" are marketed in regions where beef is not eaten.

The meat content varies from 100% including about 20% fat, to 80% or less with various additions of cereals, onion and water. Since the pattie is raw it is stored frozen.

Typical composition per 100 g raw hamburgers made from 90% meat is;- 56 g water, 15 g protein, 20 g fat (10 g saturated, 9 g monounsaturated, 1 g polyunsaturated, 100 mg cholesterol), 1.1 Ml, 600 mg sodium, 2.5 mg iron, 0.25 mg copper, 3 mg zinc, 0.04 mg B1, 0.2 mg B2, 4 mg nicotinic acid, 0.2 mg B6, 1 ug B12, 1 mg folate, 0.4 mg pantothenate.

Frying causes a small loss of water, about 5%, and a greater loss of fat depending on the method used - whether under a grill, on a heated surface or over a direct flame. Other nutrients, including protein, are proportionately increased.

## Methods of Meat Preservation Without Refrigeration

***Drying:*** Micro-organisms cannot grow unless there is sufficient moisture available to them and drying meat under conditions of natural temperatures and humidity with circulation of air and the assistance of sunshine is the oldest method of preservation (FAO 1990 c).

The free water in a food product, i.e. excluding the water bound to proteins, is termed the water activity (aw). Free water is that part that can be removed as water vapour (and is not the same as the total moisture content). "Water activity" is defined as the ratio of water vapour pressure measured in the product to the pressure of a saturated water vapour atmosphere at the same temperature.

The minimum moisture content necessary for bacterial growth varies with the type of organism. The lowest value for normal bacteria is water activity 0.91; for normal yeasts it is 0.88; for normal moulds 0.80; and for salt-tolerant (halophilic) bacteria it is 0.77.So water activity must be reduced below these levels to preserve the food.

Muscle meat of almost any kind can be dried but it is necessary to use lean meat since fat becomes rancid during the drying process. Drying involves the removal of moisture from the outer layers and the migration of moisture from the inside to the outside, so the pieces of food must be thin. The meat is cut into long thin strips or flat thin pieces and preferably salted, either dry or by dipping into salt solution, to inhibit bacterial growth and to protect from insects.

The pieces are suspended in racks in freely circulating air under hygienic conditions and protected from dirt and dust for the several days required. If the air is warm and of low humidity with relatively small temperature fluctuations between day and night the time needed will be shorter. Slow drying allows deterioration since micro-organisms can multiply in the early stages while the moisture content is still high enough. Another problem arises from the practice in developing countries of using meat from unchilled carcasses and while the temperature is still high the meat ripens rapidly so changing the flavour. At the same time there is some oxidation of the fat so further lowering the quality of the finished product.

There are a number of traditional dried products in various regions. For example biltong in South Africa, which is made from beef or antelope meat cut into strips 1 - 2 cm thick, salted, with the addition of nitrate or nitrite , spiced and dried in air for 1 - 2 weeks.

The outer layer of biltong is hard and brown with a soft, inner, red inside, and is eaten raw. It will keep for a year if stored in airtight packaging.

Typical analysis per 100 g: 11.5 g water, 1.9 g fat, 12.5 g ash, 65 g protein, 1.3 MJ (300 kcal).

Jerked beef or charque is the product used in South America which may be made from beef, llama, sheep, alpaca. The fresh meat is cut into large pieces no more than 5 cm thick, salted, pressed for several days and dried - but it still contains moisture which is allowed to drain freely from the product. It keeps for months at ambient temperatures and is resistant to insect infestation and mould growth.

Pemmican is dried meat that has been powdered or shredded and mixed with fat to form a solid product. Typical analysis per 100 g:- 3 g water, 40 g protein, 45 g fat, 2.4 MJ. Pemmican was almost a routine food taken on earlier expeditions until replaced by modern types of dried meat products.

Other traditional dried products include pastirma (Turkey, Egypt and Armenia), odka (Somalia and other countries of East Africa), qwanta (Ethiopia and East Africa) and kilishi (Nigeria and West Africa). There is a variable loss of vitamins from such products due to the long drying times which can be shortened by the use of modern drying techniques.

Such a procedure is freeze-drying which causes little or no loss of vitamins and results in products which are readily rehydrated and much closer in texture and flavour to fresh meat than the traditional dried product but calls for specialised equipment.

### *Partial Drying/Intermediate-Moisture Foods*

In dried meats the water activity is below levels needed for microbial growth so the product is shelf-stable but there will still be chemical and physical changes due to rancidity and discolouration which call for adequate packaging. Some products such as "dry" sausages and hams cannot be dried adequately without spoiling the product - they are termed "semi-dry" - so it is necessary to combine an incomplete reduction in water activity with other methods such as lowering of pH or the addition of nitrate.

In an attempt to avoid the relatively poor texture and flavour of most dried meat products a modern development is partial drying to a moisture content of 15 to 50% and then reduction of free water to the required low levels by adding humectants such as glycerol, sorbitol or other polyhydric alcohols which combine with the free water so that it cannot be used by the micro-organisms.

The meat is cut into small pieces and treated with a mild salt solution, and the humectant and an antimycotic (anti-mould agent) are added and the meat cooked to 70°C before packaging. It will keep for several months even at 38°C but there are changes in texture, colour and flavour (Lawrie 1991). Products preserved in this way are called intermediate-moisture foods and they are more succulent than dried foods but the humectants spoil the palatability and the process has been limited to animal foods in industrialised communities and for military purposes.

### Canning

Micro-organisms can be completely destroyed by heat (sterilisation) but a sterile product can be readily recontaminated unless it is protected. This is achieved by heating in an air-tight can or bottle, or, more recently, in a heat-resistant or aluminium foil-laminated plastic pouch. Sausages can be filled into retortable synthetic casings sealed with aluminium clips.

The procedure is to seal the food into the container and then heat it under pressure in an autoclave (retort) to the required temperature for the required length of time and to cool rapidly to avoid overheating. Overheating results in too soft a consistency and a burnt taste. It is not always possible to destroy all the organisms without excessive heat which would spoil the product so the objective is to destroy the greater proportion of the organisms when the remaining few pose no hazard so long as the container is cooled rapidly and stored below 20-25°C. This condition is termed "commercially sterile". The established standard is equivalent to a reduction in the number of micro-organisms by a factor of 10 to the power of 12 so it is clear that the higher the initial load of organisms the more will survive a standard heat treatment.

The intensity of heat treatment necessary depends on the nature of the product, its pH, and the amount of salt and other curing agents present as well as on the bacterial load. The time required at a given temperature will vary with the rate of heat penetration to the centre and so with the size of the container.

The intensity of heat treatment is defined in physical terms called F-value, which means that the product received heat treatment with the same effect on micro-organisms as exposure to a temperature of 121°C for 1 minute. The standard is based on the time required at a temperature of 121°C to destroy all spores of Clostridium botulinum, the most dangerous of all toxin-producing organisms. This is termed "the botulinum cook" and such treatment destroys practically all spoilage and other organisms. It takes 2.45 minutes at 121°C to destroy all C. botulinum spores; this is an F-value of 2.45. Spores of other organisms are less or more heat-resistant. F-value 1 is the lethal effect on micro-organisms after 1 minute at 121°C; F-value 2 (3,4) is the lethal effect after 2 (3,4) minutes. At temperatures greater than 121°C a shorter time is needed to achieve the F-value of 1, thus at 130°C the time is 0.13 min. Correspondingly the time is longer at lower temperatures, thus at 115°C the time is 4 min at 105°C it is 40 min.

These conditions apply to foods of low acidity (pH above a value of 5) and medium acidity (pH 4.5 - 5); with more acid foods the spores of micro-organisms are less heatresistant. Meat products are mostly low-acid, while meat and vegetable mixtures are medium -acid. In practice once the F-value has been determined for a batch of food according to the size of the container the heat treatment required to treat subsequent batches is the same. Generally it has been shown that F-value 4 will usually ensure commercial sterility. Larger canned products may require F-values up to 20-25 owing to the longer periods required for heat penetration.

A fully-treated product of this type will keep for up to 4 years at ambient temperatures but even fully-preserved meat can contain a very heat-stable spore former, Clostridium sporogenes, which poses a hazard only when stored under extreme climatic conditions, namely at temperatures above about 40°C If canned meat is to be stored under such conditions then it must be treated more intensively, F-value 12 or more ("tropical preservation") and then has a shelf life up to 4 years.

Virtually every type of meat product made from chopped, cured meat can be canned, as well as stewed meat, dishes in jelly, soups with meat ingredients, and pastas and sausages in brine.

Products such as luncheon meats, liver sausage, blood sausage and jellied products are adversely affected by high temperatures and are "three-quarters preserved" at F-values 0.6 to 0.8. The temperature reached at the centre of the pack is between 108 and 112°C and the product is stable for up to 1 year if stored at temperatures no higher than 15°C.

Cooked preserved products are simply boiled until the central temperature reaches near to 100°C and they can be stored (protected from contamination) for 1 year at temperatures no higher than 10°C.

Smaller size containers are most suitable for meat products because heat penetration is mostly by conduction so larger containers would require severe heat treatment involving overcooking. Large pieces of meat products such as hams, shoulders, etc., are pasteurised.

Pasteurisation is a more gentle process intended to destroy only pathogenic organisms and the treatment limits the central temperature to about 80°C (Fvalue almost zero). This destroys only vegetative cells and refrigeration is necessary to prevent germination of spores. Pasteurised products must be stored between 2 and 4°C when they have a shelf life up to 6 months.

The temperatures quoted must be reached in the centre of the pack to ensure that the entire contents are adequately heated but protein and fat are poor heat conductors. If there is enough liquid in the can, such as meat cooked in gravy, Frankfurters in brine, or through release of liquid from the meat and liquefaction of the fat, heat can penetrate by convection as well as conduction if the can is rotated during the process. This allows a shorter heating time with less damage to flavour, texture and nutrients, and the outer layer of the food is not overheated.

Canning operations must be performed only by fully-trained personnel (FAO 1990c; Hershom and Hulland 1980).

### High-temperature Short-time Processing (HTST)

Since the effect of heat in speeding up biological reactions (in this instance destruction of micro-organisms) is greater than the acceleration of chemical reactions (in this instance damage to protein and other nutrients) heating to a higher temperature for a shorter time is an effective means of preservation. Sterilisationis achieved in a shorter time with less damage to the product. The process is termed high-temperature short-time heating (HTST) and has been particularly applied to milk but can be applied to meat if there is sufficient liquid present to allow mixing of the contents by rotating the cans in the autoclave. The cans must be cooled immediately after the temperature of sterilisation has been reached to avoid overheating.

### Nutritional Damage

Most of the investigations of nutritional changes in canning have involved fullypreserved foods. There are large differences reported in the literature due to differences in raw materials and conditions of processing.

Thiamin is the most labile of the vitamins and reported losses range between 20 and 40%. This is heat destruction since any water-soluble nutrient that is leached out of the meat will be retained in the can and is usually consumed together with the meat. Losses of niacin and riboflavin are about 10%, 20% of biotin, 20-30% of pantothenate. If more accurate figures are required they would need to be determined on the specific procedure in the factory in question on the particular product.

In principle losses in canning are somewhat higher than in wet cooking since the temperature is higher, and for the parallel reason less than losses in dry cooking methods such as roasting, grilling and frying where the temperature at the surface can be between 180 and 350°C So far as proteins are concerned there is some small reduction in biological value due to reduced availability of methionine and cystine. Semi-preserved meat products heated to temperatures not exceeding 100°C do not suffer this damage.

The heat applied causes partial hydrolysis of the collagen so that tough meat, rich in collagen, is rendered more palatable by being canned. During storage after canning there can be a loss of thiamin of as much as 30% depending on the length of time, and at higher storage temperatures there can be a reduction in protein quality.

### Fat Embedding

A traditional process that is parallel to canning is that of cooking the meat in a vessel that can be sealed under a layer of melted fat and so protected from recontamination. An example of such a product is mixiria of the Amazon region where the meat is roasted, sliced and sealed in jars. The layer of fat not only protects the meat from contamination but excludes oxygen, however, organisms can survive so the method is not dependable.

The process of fat embedding was tested by the Australian meat trade in the 19th century - beef was packed into barrels and covered with fat heated to 150°C - but superseded by refrigeration.

It is used to a very limited extent in some industrialised countries, in particular for a product called "potted shrimps", which have been cooked in butter and sealed into jars.

### Ionising Irradiation

Micro-organisms can be destroyed by subjecting a food to ionising radiation produced from radioactive or electromagnetic sources. High doses (50 kiloGrays - kGy) are required for sterilisation of meat while recommendations of WHO and legislation in most, if not all, countries limit the dose at present to 10 kGy, simply because safety has been established up to this level.

The 10 kGy dose does not sterilise the product but substantially reduces the bacterial load and is effective in destroying many pathogens including Salmonellae. A dose of 2 - 5 kGy will extend the shelf life of poultry stored at 1-3°C by 8-14 days. Irradiation is of no value as a means of "cleaning up". heavily contaminated food since it would still carry a considerable microbial load, nor does it destroy toxins once they have been produced. Indeed, in some countries where irradiation is permitted microbiological standards are specified for foods to be treated. Irradiation of spices to be used for meat products has proved to be an effective way of lowering their microbial content and so increasing the shelf life of the products.

Although the preservation of food by irradiation has been intensively studied for many years its commercial application is still in its infancy. There are many problems involved since the process calls for heavy investment in factory plant and is regarded with some suspicion by consumers. Moreover irradiation does not destroy enzymes so the meat softens during storage.

Since the radiation penetrates into the product it also penetrates packaging so the food can be protected from recontamination by adequate wrapping before irradiation.

In general irradiation has some deleterious effect on vitamins but the amount of damage is not considered nutritionally significant (Codex 1984A).

## Recent Developments in Meat Processing

### *Mechanically Recovered Meat (MRM)*

Also termed mechanically deboned and mechanically separated meat.

When bones are trimmed of adhering meat by hand some of the meat is left attached to the bones. It is said that some 2 million tonnes of meat world -wide are discarded with the bone. This can be recovered by a mechanical scraping process in which bones with adhering meat are generally broken into small pieces and the soft components, meat, fat etc., forced through perforated screens and recovered as a finely comminuted product consisting of a gel of meat, fat and sometimes marrow from the inside of the bones together with fine particles of bone. The bone content should be reduced to a minimum; it has been recommended that the calcium content of MRM expressed on a dry matter basis should not exceed 1.5% (Codex 1983).

MRM can increase the overall profitability of meat production not only by increasing meat yield but also by providing a product with emulsifying properties useful in preparing comminuted meat products. Generally MRM is less acceptable than ordinary muscle meat because it is darker in colour due to its high content of haeme protein, and more readily oxidised because of the high fat content.

Because of its physical state and the longer process MRM is subject to heavy microbial contamination; if not used immediately it needs freezing to prevent decomposition.

### *Reformed Meat Products*

Apart from the high regard for certain cuts of meat in some areas there is a growing demand for lean rather than fatty cuts. The high cost of such products has led to the use of other, less expensive cuts of meat which are higher in content of connective tissue but can have some of this tissue, together with skin and gristle, removed in a process of comminution and reformation.

The pieces of meat are treated in a tumbling device with brine which penetrates the tissues and extracts some of the myofibrillar protein to form a gel.

At a low temperature the product can be compressed into thin flakes which bind together when the meat is heated, sometimes with the addition of binding agents such as wheat gluten, egg albumin and soya; the final product can be shaped or sliced.

The nutritive value of reformed meat products depends on the tissue used and the added ingredients, and approximates to that of comminuted products.

### *Protein Extraction*

A recently discovered but highly technical process for recovering edible foodstuffs from by-products of low acceptability (such as lungs and stomachs) is extraction of the proteins.

The procedure is similar to that of preparing textured vegetable proteins usually made to simulate meat. The proteins are extracted in alkali and reprecipitated from solution by acidifying. If the alkaline solution is forced through fine tubes (spinnerets) into the acid solution it is precipitated in the form of long fibres which can be pressed and texturised to a solid form. High temperature extrusion is also used. Other materials such as soya can be incorporated.

There is an added advantage in that the number of micro-organisms is greatly reduced during the process.

A recent method of using low quality meats is the application of the Japanese fish surimi process to meat and poultry by-products. This consists of mechanically deboning and washing to remove water-soluble proteins, enzymes, minerals and fat to leave a product that is stable and with a high concentration of salt-soluble myofibrillar proteins that have the capacity to form strong elastic gels when gently heated.

Fish surimi is used to make a range of products by extrusion and shaping and it seems possible to apply the procedure to meat. Low-quality meat is rich in fat and the process is intended to remove the greater part of the fat and to make novel meat products or to add to traditional meat products such as sausages, sliced cooked meat and burgers or other foods.

## Meat Consumption Role of Meat in the Diets

### *Meat Consumption*

Meat consumption is based largely on availability, price and tradition. Meat production is a very complex operation depending not only on demand (which is usually based on price and income) but on many social and economic influences such as official policy, price support mechanisms, and interrelations such as the interaction between beef and milk production, the availability of animal feedstuffs and competition for food between man and animals.

It is difficult to make accurate comparisons of meat consumption between countries because different methods are used to estimate consumption. Figures may be derived from total supplies available at wholesale level, or from records of household purchases, with or without estimates of what is consumed away from home; the estimate of waste, both in preparation of the food and by the individual adds to the uncertainty.

Some national estimates fail to include imports, and some surveys include the weight of non-meat components of the products (for example the amount of meat in a product can range between 100% in some burgers to 10% in some pizzas). FAO Food Balance Sheets are prepared from figures for production, imports, stock changes and exports with allowances for feed, processing and "other uses" and the same methods are applied to all regions so that the figures.

Up to a certain level of income the amount of meat eaten varies with income, in the relatively affluent western world where the proportion of available income spent on food has been steadily falling over the past generation, there is now little if any difference between the amounts of meat eaten by the different income groups. This contrasts with the Third World countries (FAO 1990a).

The amount of meat consumed in different countries varies enormously with social, economic and political influences, religious beliefs and geographical differences. It is very large in meat-producing areas such as Uruguay, Argentina, Australia and New Zealand, at 300 g per head per day compared with an average of 10 g in India Indonesia and Sri Lanka, the contrast between total meat supplies in developed and developing countries, allowing for exports, imports and stock changes, and that the production per capita in the former is five times as much as in the developing countries. These tables also show the relative size of production of the different types of animals involved.

### Role of Meat in the Diet of Developed and Developing Countries

Meat is held in high esteem in most communities. It has prestige value, it is often regarded as the central food round which meals are planned, various types of meat are sometimes made the basis of festive and celebratory occasions, and from the popular as well as the scientific point of view, it is regarded as a food of high nutritive value.

While it is clear that meat is not essential in the diet, as witness the large number of vegetarians who have a nutritionally adequate diet, the inclusion of animal products makes it easier to ensure a good diet.

There is a marked difference at the present time in attitudes towards meat between the people of the developing and industrialised communities In the former where meat is in short supply it can be taken as a measure of the nutritional quality of the diet as a whole. Where a typical diet is heavily dependent on one type of cereal or root crop, meat, even in small amounts, complements the staple food. It provides a relatively rich source of wellabsorbed iron and also improves the absorption of iron from other foods, its amino acid composition complements that of many plant foods, and it is a concentrated source of B vitamins, including vitamin B12 which is absent from plant foods. Consequently there is pressure to increase the availability of meat products.

In the industrialised countries where food of all kinds is plentiful and cheap there is concern, whether or not misplaced, about the potentially harmful effects of a high intake of saturated fat from animal foods (discussed later), emphasis on continuous development of regulations dealing with hygiene in slaughter houses and during subsequent handling, concern about hormones administered to cattle, what is perceived as excessive addition of water to some processed products - concerns that can scarcely be afforded in developing countries when balanced against food supplies.

With increasing mechanisation in industrialised communities the steady fall in human energy expenditure and consequently in per capita food consumption poses a potential problem in achieving an adequate intake of nutrients even where there is an abundance of food available. With the variety of food available a diet of 8 Ml (2000 kcal) or more per day is likely to supply enough of all the nutrients, but when the intake is 6.5 to 7 MJ (1600-1800 kcal) per day the consumer needs to make an informed choice of foods to ensure an adequate intake of nutrients.

In western Europe the daily average energy intake of women is about 6.5 Ml and that of men 8 MJ (excluding alcohol) and there are reports of biochemical signs of deficiencies of several B vitamins and iron. It is not clear whether this is accompanied by functional defects.

In industrialised countries there have been slow but continuous changes over the years in the relative amounts of different types of meat consumed (beef, pork, lamb, poultry) depending partly on price and influenced by fashion, advertising, etc. In more recent years health aspects, more correctly, perceived health aspects, have become a factor.

The concerns about public health in industrialised countries where coronary heart disease and other "diseases of affluence are common have led to recommendations to the public to modify their diet, popularised as Dietary Guidelines. These particularly recommend a reduction in fat consumption, especially saturated fatty acids and consequently, even if incorrectly, in red meat (discussed later). This has led in some sections of their populations to a relative increase in the consumption of poultry and fish at the expense of red meat.

In addition there is concern, whether or not misplaced, about the presence in meat of pesticides, residues of hormones and growth promoters used to increase yields, and concern about human diseases thought to be transmitted by beef, together with an increase, for many reasons, in vegetarianism.

## Meat as a Source of Protein

### Human Protein Requirements

Human requirements for protein have been thoroughly investigated over the years (FAD/WHO 1985) and are currently estimated to be 55 g per day for adult man and 45 g for woman. (There is a higher requirement in various disease states and conditions of stress).

These amounts refer to protein of what is termed "good quality" and highly digestible, otherwise the amount ingested must be increased proportionately to compensate for lower quality and lower digestibility.

### Protein Quality

The quality of a protein is a measure of its ability to satisfy human requirements for the amino acids. All proteins, both dietary and tissue proteins, consist of two groups of amino acids - those that must be ingested ready-made, i.e. are essential in the diet, and those that can be synthesised in the body in adequate amounts from the essential amino acids. Eight of the 20 food amino acids are essential for adults and ten for children.

The quality of dietary protein can be measured in various ways (FAD/WHO 1991) but basically it is the ratio of the available amino acids in the food or diet compared with needs. In the earlier literature this was expressed on a percentage scale but with the adoption of the S.I. system of nomenclature it is expressed as a ratio. Thus a ratio of 1.0 (100 per cent) means that the amino acids available from the dietary proteins are in the exact proportions needed to satisfy human needs; a ratio of 05 means that the amount of one (or more) of the essential amino acids present is only half of that required. If one essential amino acid is completely absent (a circumstance that can occur only experimentally with isolated proteins since any food, let alone a whole diet, consists of a mixture of many proteins) the protein quality would be zero.

There is a popular impression, originating at one time from nutrition textbooks, that the qualities of proteins from animal sources are greatly superior to those from plant sources. This is true only to the extent that many animal sources have Net Protein Utilisation,

NPU, (a measure of the usefulness of the protein to the body) around 0.75 while that of many, but not all plant foods is 0.5-0.6. However, after infancy people consume a wide variety of proteins from different foods and a shortfall in any essential amino acids in one food is usually made good, at least in part, by a relative surplus from another food - this is termed complementation. As a result the protein quality of whole diets even in developing countries rarely falls below NPU of 0.7, a value that can be compared with the average of 0.8 in industrialised countries (FAD/WHO 1985).

The value of meat in this respect is that it is a relatively concentrated source of protein, of high quality (NPU 0.75-0.8), highly digestible, about 0.95 compared with 0.8-0.9 for many plant foods, and it supplies a relative surplus of one essential amino acid, lysine which is in relatively short supply in most cereals.

### Effect of Cooking on Protein Quality

Apart from the inherent quality of the various proteins a reduction in quality takes place if there is damage to amino acids when the food is cooked. At a temperature below 100°C when proteins are coagulated, there is no change in nutritional quality.

The first changes take place when food is heated to temperatures around 100°C in the presence of moisture and reducing sugars, present naturally or added to the food. There is a chemical reaction between part of one essential amino acid, lysine and a sugar to form a bond that cannot be broken during digestion, and so part of the lysine is rendered unavailable.

When proteins are analysed in order to determine their amino acid composition the procedure involves a preliminary hydrolysis with strong acid which does break the lysinesugar bond, so chemical analysis does not reveal this type of damage and special methods are needed. At a higher temperature or with more prolonged heating, the lysine in the food protein can react with other chemical groupings within the protein itself and more becomes unavailable. In addition the sulphur amino acids (cystine which is not essential and methionine which is) are rendered partly unavailable.

The lysine-sugar reaction results in a brown-coloured compound (the so-called browning or Maillard reaction) which produces an attractive flavour in food and is the main cause of the colour of bread crust and roast meat. While such severe heating reduces the amount of lysine available in these foods the loss is nutritionally insignificant since it affects only a very small fraction of the total amount present. At the temperature needed to cook meat there is little loss of available lysine or the sulphur amino acids but there can be some loss if the meat is heated together with reducing substances, as may be present when meat is canned with the addition of starch-containing gravy or other ingredients.

Overall the damage to protein caused by cooking is of little practical significance and it can be argued that if there is meat in the diet it is likely that the quantity of protein would compensate for any shortfall in quality.

The nutritional quality of the proteins of meat rich in connective tissue is low since collagen and elastin are poor in the sulphur amino acids - there is only 0.8 g of each per 100 g of total protein compared with values of 2.6 and 1.3 of each respectively in "good meat. Meat is tough to eat when it is rich in connective tissue and such meat is often used for canning since the relatively high temperature involved in the sterilisation process partly hydrolyses the collagen so making the product more palatable. However, it still results in a product with NPU as low as 0.5 compared with a value of 0.75 - 0.8 for good quality meat (Bender and Zia 1976).

### *Adequacy of Dietary Protein*

The protein requirement of an individual is defined as the lowest level of protein intake that will balance the loss of nitrogen from the body in persons maintaining energy balance at modest levels of physical activity (FAD/WHO 1985). The "requirement" must allow for desirable rates of deposition of protein during growth and pregnancy. When energy intake is inadequate some of the dietary protein is diverted from tissue synthesis to supply energy for general physical activity - this occurs at times of food shortage and also in disease states where food is incompletely absorbed and utilised.

A diet adequate in energy is almost always adequate in protein - both in quantity and quality. For example, an adult needs an amount of protein that is equivalent to 7 - 8% of the total energy intake, and since most cereals contain 8 - 12% protein even a diet composed entirely of cereal would, if enough were available and could be consumed to satisfy energy needs, satisfy protein needs at the same time. Growing children and pregnant and nursing mothers have higher protein requirements as do people suffering from infections, intestinal parasites and conditions in which protein catabolism is enhanced. During the stress that accompanies fevers, broken bones, burns and other traumas there is considerable loss of protein from the tissues which has to be restored during convalescence and so high intakes of protein are needed at this time together with an adequate intake of energy.

The digestibility of the proteins of various diets varies considerably. For example, the digestibility of typical Western diets and Chinese diets is 0.95 (i.e. 95% digested). That of the Indian rice diet and Brazilian mixed diet is 0.8 (FAD/WHO 1985). Digestibility is high in diets that include milk and meat and low when maize and beans predominate.

An increase in the amount of protein eaten beyond "requirement" figures compensates for any shortfall in digestibility and protein quality.

## Meat as a Source of Vitamins and Minerals

Meat and meat products are important sources of all the B-complex vitamins including thiamin, riboflavin, niacin, biotin, vitamins B6 and B12, pantothenic acid and folacin. The last two are especially abundant in liver which, together with certain other organs is rich in vitamin A and supplies appreciable amounts of vitamins D, E and K.

Meats are excellent sources of some of the minerals, such as iron, copper, zinc and manganese, and play an important role in the prevention of zinc deficiency, and particularly of iron deficiency which is widespread.

***Meat Iron:*** The amount of iron absorbed from the diet depends on a variety of factors including its chemical form, the simultaneous presence of other food ingredients that can enhance or inhibit absorption, and various physiological factors of the individual including his/her iron status. Overall, in setting Recommended Daily Intakes of nutrients the proportion of iron absorbed from a mixed diet is usually taken as 10%.

Half of the iron in meat is present as haeme iron (in haemoglobin). This is well absorbed, about 15-35%, a figure that can be contrasted with other forms of iron, such as that from plant foods, at 1-10%.

Not only is the iron of meat well absorbed but it enhances the absorption of iron from other sources - e.g. the addition of meat to a legume/cereal diet can double the amount of iron absorbed and so contribute significantly to the prevention of anaemia, which is so widespread in developing countries.

Zinc is present in all tissues of the body and is a component of more than fifty enzymes.

Meat is the richest source of zinc in the diet and supplies one third to one half of the total zinc intake of meat-eaters. A dietary deficiency is uncommon but has been found in adolescent boys in the Middle East eating a poor diet based largely on unleavened bread.

### Health Concerns Associated with the Consumption of Meat

***Coronary or Ischaemic Heart Disease:*** A major cause of death in some parts of the industrialised world is coronary heart disease (CHD) and saturated fatty acids have been implicated as an important dietary risk factor. Since about a quarter of the saturated fatty acids in the diet is supplied by meat fat, the consumption of meat itself has come under fire.

The first stage of development of the disease is a narrowing of the coronary arteries by deposition of a complex fatty mixture on the walls - a process termed atherosclerosis. The fatal stage is the formation of a blood clot that blocks the narrowed artery thrombosis. Even if the thrombosis is not fatal the reduced blood flow to the heart muscle deprives it of oxygen and can lead to extensive damage - myocardial infarction.

Despite many years of intensive investigation the real cause of CHD is not known but a large number of what are termed risk factors have been identified, including a family history of CHD, smoking, lack of exercise, various types of stress and certain disease states together with a number of dietary factors. The saturated fatty acids, myristic and palmitic, have been established as the most important of the dietary risk factors in coronary heart disease.

There are three types of lipoproteins (protein-lipid complexes) in the blood; low density lipoproteins (LDL) in which 46% of the molecule is cholesterol; high density lipoproteins (HDL) which include 20% as cholesterol; and very low density lipoproteins (VLDL) which have 8% cholesterol. High levels of total blood cholesterol are associated with the incidence of CHD and high intakes of saturated fatty acids elevate blood cholesterol levels: hence the association between dietary saturated fatty acids and CHD. It is the LDL that appear to be the main problem and HDL appear to be protective.

This lipid hypothesis of causation of CHD has led to the adoption in many countries of dietary guidelines which, among other objectives, are intended to reduce the intake of saturated fatty acids as compared with unsaturated fatty acids and so reduce blood levels of LDL.

## Types of Fatty Acids

### *Saturated Fatty Acids (SFA)*

Two of the saturated fatty acids, myristic and palmitic acids, appear to be the principal dietary factors that increase the blood cholesterol and do so by increasing LDL The other main SFA in the diet, stearic acid, does not have the same effect (apparently because it is converted to oleic acid which is monounsaturated); fatty acids of shorter chain length appear to have no effect.

In order to explain the terms saturated and unsaturated fatty acids to the consumer, SFA have been equated with animal fats so meat fat is perceived as being saturated, but, in fact, this is only relative.

For example pork lard is 40% SFA, beef tallow is 43-50% SFA, depending on the part of the body from which it is derived. These figures can be compared with 20 - 25% SFA in vegetable oils which are perceived as unsaturated. That except for lamb fat the proportion of SFA is about 40% or less. In four of the six samples of meat listed there is a higher proportion of monounsaturates than SFA.

This perception of meat fat as being saturated has led to the belief that meat, particularly red meat, should be avoided. In fact it has been shown that a reduction of total fat intake while still including in the diet 180 g of lean meat containing 8.5% fat can result in a reduction in blood cholesterol levels (Watts et al 1988).

The relation between diet and coronary heart disease is not only a subject of considerable misunderstanding in the minds of consumers but also a subject of some controversy among medical scientists.

### *Monounsaturated Fatty Acids (MUFA)*

The fatty acid of main interest is oleic acid (plentiful in olive, rape seed and higholeic safflower oils). The relatively high intake of olive oil and consequently the proportionately low intake of SFA are believed to be important dietary factors in the low incidence of CHD in Mediterranean countries compared with northern Europe.

It is not clear whether oleic acid confers direct protection or simply replaces SFA in the diet. The contribution of meat fat to the intake of MUFA.

### *Polyunsaturated Fatty Acids (PUFA)*

These are fatty acids with between 2 and 6 double bonds and long carbon chains of 18 to 22 carbon atoms. Linoleic acid (18 carbon atoms and 2 double bonds) and linolenic acid (18 carbon atoms and 3 double bonds) are plentiful in many vegetable oils. The very long chain fatty acids, eicosatetraenoic (20C, 4 double bonds) and docosapentaenoic (22 C, 5 double bonds) are plentiful in fish oils and smaller amounts are present in some meat fats.

These very long chain PUFA appear to offer direct protection against "heart disease", particularly against thrombosis, but it is not clear whether the other PUFA in the diet (from vegetable oils) offer protection or simply displace SFA. Consequently it is often recommended that vegetable oils (rich in PUFA) should not simply be added to a diet but should be used to replace other fats when there is a need for fat in formulating food products.

Linoleic and linolenic acids are essential in the diet (they were at one time termed vitamin F) and the very long chain FA are formed from them in the body. It is possible that the rate of their formation may not be adequate under all circumstances and so there may be benefit from consuming some of these very long chain PUFA ready-made in the diet.

### *Trans Fatty Acids*

PUFA exist in nature in two structural forms, termed cis and bans forms. It is the cis forms that are used in the production of fatty products such as special margarines. The other forms, bans, are formed when oils are hydrogenated to make hard fats for some margarines, and small amounts are found in the fats of ruminants where they are formed by bacterial hydrogenation in the rumen.

Experimentally bans fatty acids have been shown to have an adverse effect on both LDL and HDL and so are considered potentially harmful. When calculating the ratio of PUFA and SFA in diets, the bans fatty acids are often included with SFA.

## Cholesterol

Cholesterol is a fatty compound involved in the transport of fat in the blood stream and is also part of the structure of cell membranes of tissues of the body. It is not a dietary essential since adequate amounts are synthesised in the body from other dietary ingredients. Confusion has arisen between the terms blood cholesterol and dietary cholesterol. For most individuals dietary cholesterol has little or no effect on blood cholesterol levels because reduced synthesis in the body compensates for increased dietary intake. However, there are individuals who are sensitive to dietary cholesterol (Reiser and Shorland 1990) and most authorities advise a general reduction in cholesterol intake for everyone.

Meat supplies about one third of the dietary cholesterol in many western diets with the remainder from eggs and dairy products. Since all these foods are valuable sources of nutrients there could be some nutritional risk in restricting their intake.

In addition to playing an important role in CHD dietary saturated fats have been implicated in hypertension, stroke, diabetes and certain forms of cancer, so all dietary guidelines include recommendations to reduce total fat intake and especially that of saturated fats.

Some 20 national authorities have issued dietary guidelines which differ mainly in the amounts of the various foods advised (James 1988). Generally it is recommended that total fat should be reduced to 20-30% of the total energy intake, with not more than 10% from saturates, 10-15% from MUFA and with PUFA at 3% or more; this results in a P/S ratio of 1.0.

Most authorities, but not all, recommend a reduction in dietary cholesterol to around 300 mg or less per day.

## Poultry Meat Versus Read Meat

Dietary guidelines sometimes include advice to substitute, at least in part, chicken for red meat. Chicken meat including its skin contains about the same amount of fat as does medium-fat red meat, 20%; it is necessary to remove the skin with the adhering subcutaneous fat, to reduce the fat content to around 5% - which is no lower than the figure for lean meat.

However, chicken flesh has less saturated fatty acids (33% of the total) and more PUFA (14%) than lean meat with 45% and 4% respectively.

Duck flesh is very fat, containing about 10% fat - 45% when the skin and subcutaneous fat are included; only 27% of duck fat is saturated. Meat from game birds, grouse, partridge, pheasant and pigeon, contains about 5, 7, 9, and 13% fat respectively, of which about one quarter is saturated.

Apart from differences in the amounts and types of fatty acids in the various kinds of meat, poultry and game their nutrient compositions are similar.

### Toxic Compounds Formed During Processing and Cooking

While cooking is necessary to develop the desirable flavours in meat (as well as to destroy harmful organisms) the oxidation of fats, especially at frying temperatures, can give rise to compounds that decompose to aldehydes, esters, alcohols and short chain carboxylic acids with undesirable flavours. Meats are particularly susceptible because of the unsaturated lipids present which are more readily oxidised and because of catalysis by haeme and non-haeme iron.

The more PUFA present the greater the likelihood of oxidation, and pork (3.6 g PUFA/100 g when grilled), duck (meat and skin, cooked, 3.5 g) and chicken (roast meat and skin, 2.5 g) are the most susceptible. Other types of meat are less susceptible, e.g. Iamb (grilled cutlets, 1.5 g PUFA), turkey (meat with skin, 1.3), and beef (fried steak, 0.6 g per 100 g).

The adverse effect of these oxidation products on eating quality is well recognised but more recently it has been suggested that some of them may be carcinogenic, and also may be involved in the ageing process and CHD. However, it is possible or even likely that the unpleasant flavours would cause rejection of the food at levels below harmful ranges.

Cholesterol can also be oxidised and the oxidation product has been suggested as a possible factor in CHD (Addis 1986).

### Carcinogens

A number of epidemiological studies have suggested a link between the intake of animal protein and predisposition to cancers at various sites -pancreas, breast, colon, prostate and endometrium - but there are many contradictory reports. A summary of eleven case-controlled studies of colon cancer, three of stomach cancer and one of breast cancer concluded that the available data do not provide convincing evidence that removal of meat from the diet would substantially reduce the cancer risk (Phillips et al 1983; Kritchevsky 1990).

The products of pyrolysis of organic material (by overheating and charring), polycyclic hydrocarbons, are believed to be carcinogenic. The most thoroughly investigated of these is 3,4-benzpyrene which is formed on the surface of barbecued and broiled (grilled) and smoked meat products (including broiled fish and roasted coffee). The main source of these compounds is the flame itself, especially from charcoal, and indirect cooking where the flame is not in contact with the food greatly reduced the amount present.

### Nitrosamines

Nitrites, used in curing salts can react with amines commonly present in food, to form nitrosamines.

These have been shown to be carcinogenic in all species of animals examined but it is not clear, despite years of intensive research, whether the amounts present in cured meats affect human beings. The problem is particularly difficult because nitrosamines have been found in human gastric juice, possibly formed from nitrites and amines naturally present in the diet. As a precaution, legally enforced in some countries, there is a tendency to reduce the amount of nitrite used in the curing mixture and to add vitamin C which inhibits the formation of nitrosamines.

Erythorbic acid and tocopherol are also effective in reducing nitrosamine formation. The problem is complex since the process of curing is designed to prevent the growth of Clostridium botulinum which is responsible for botulism, and the risk of botulism is increased if the concentration of nitrate-nitrite is reduced too far. (Moreover, cigarettes contribute far greater amounts of nitrosamines, up to one hundred times as much as cured meats).

### Other Potential Problems

***Bovine Spongiform Encephalopathies (BSE):*** There is a group of diseases called prion disease, also known as spongiform encephalopathies or transmissable dementias, which include some very rare human diseases, scrapie in animals and BSE. It is not clear whether these all represent the same disease but they have in common the presence of an aberrant form of a normal cell protein called prion protein.

In some countries there have been recent outbreaks of BSE in cattle with the suspicion that it might be transmitted to human beings through affected meat.

This is difficult to prove or disprove and the risk may be remote but it has added to other popular suspicions about meat and may be partly responsible for the reduction in beef consumption in some countries.

***Excessive Amounts of Vitamin A in Liver:*** There are reports in the scientific literature of harmful effects of acute and chronic excessive intakes of vitamin A, mostly from pharmaceutical preparations. Recently, however, concern has been expressed at unusually high levels of vitamin A found in some, few, samples of animal liver, which, if eaten during the early stages of pregnancy, might possibly affect the human foetus.

Residues of Drugs. Pesticides. etc.

Residues of drugs, pesticides and agricultural chemicals can be found in small amounts in meat and meat products. Pesticides, for example, may be applied specifically to the animals to control insects or intestinal parasites but may also be present in meat as a result of exposure of the animals to chemicals used on buildings, grazing areas and crops. While there is no clear evidence that these small amounts cause harm to the consumer they are perceived as a risk. For this reason there is widespread legislation to test for and control a range of chemical substances that may be present in meat (Codex 1991A).

The problem is complicated because several hundred substances are used to treat animals, to preserve animal health and to improve animal production. These include antimicrobial agents, beta-adrenoreceptor blocking agents (used to prevent sudden death in pigs due to stress during transport) anti-helminthics, tranquillizers, anti-coccidial agents, vasodilators and anaesthetics.

Potential safety problems arise from the possibility of residues of these drugs and their metabolites remaining in the tissues (and milk) consumed by human beings. Some tranquillizers, for example, are used in pigs in the immediate pre-slaughter period when there is no time for their removal through the normal metabolic processes . They can persist in the human body so that repeated intakes could possibly result in accumulation of the drugs.

In order to protect the consumers from such risks the Codex Alimentarius Commission publishes Draft Codes of Practice for control of the use of veterinary drugs (Codex 1991A). These provide guidelines for the prescription, application, distribution and control of drugs.

Where there is sufficient scientific information available about the drug in question the Codex Commission defines the following:- Acceptable Daily Intake (ADI) as a measure of the amount of a veterinary drug, expressed on a body weight basis, that can be ingested over a life-time without appreciable health risk (the same term and definition as used for food additives). This is set at one hundredth of the maximum no-observed-effect level (NOEL) determined in experimental animals, on the assumption that human beings may be ten times as sensitive as the test animals used to determine NOEL and that there may be a tenfold range of sensitivity within the human population. When data are incomplete the safety factor may be set at a much higher multiple.

The maximum amount of residue of a drug - maximum residue limit (MRL) - is the maximum concentration per kg fresh weight of food that is recommended by the Codex Commission as being legally acceptable. This is based on the amount considered to be without any toxicological hazard to human health and takes account of other relevant public health risks as well as food technological aspects.

A point is made in the 1991 report that the principal problem is not only the safety of the substances and their residues but the public perception of their safety. There is no doubt that administration of drugs to animals (and birds) is always a potential risk to human health and so there is a need to control the use of these drugs and to measure the extent of any residues left in the food intended for human beings.

## Conclusion

Meat is not an essential part of the diet but without animal products it is necessary to have some reasonable knowledge of nutrition in order to select an adequate diet. Even small quantities of animal products supplement and complement a diet based on plant foods so that it is nutritionally adequate, whether or not there is informed selection of foods.

Side by side with these known benefits of including meat and meat products in the diet are problems associated with excessive intakes of saturated fats, risks of food poisoning from improperly processed products, residues of chemicals used in agriculture and animal production and other potentially adverse aspects discussed. Within these concepts is the major problem of meat production under conditions that avoid food poisoning and satisfy the economic demands of profitability with the traditional, cultural and religious concerns of the community in question.

There is a steadily increasing demand for meat in the developing countries which can be satisfied by increased domestic consumption and/or increased imports. It is thought that the major increase in domestic production will come from small producers rather than from creating large production units but these lack the essential facilities for producing safe and wholesome products.

If there is to be a significant increase in meat production it will require clear policy decisions with the necessary financial, legislative and technical support. There is considerable potential for increased supplies through better management, selection of animals, avoidance of waste and making use of indigenous species.

If exports are to be considered then attention has to be paid to the strict hygienic and safety requirements involved, whatever the domestic market might tolerate.

## Poultry Farming

Poultry farming is the raising of domesticated birds such as chickens, turkeys, ducks, and geese, for the purpose of farming meat or eggs for food. Poultry are farmed in great numbers with chickens being the most numerous. More than 50 billion chickens are raised annually as a source of food, for both their meat and their eggs. Chickens raised for eggs are usually called layers while chickens raised for meat are often called broilers. In total, the UK alone consumes over 29 million eggs per day. In the US, the national organization overseeing poultry production is the Food and Drug Administration (FDA). In the UK, the national organisation is the Department for Environment, Food and Rural Affairs (Defra).

### *Intensive and Alternative Poultry Farming*

According to the Worldwatch Institute, 74 percent of the world's poultry meat, and 68 percent of eggs are produced in ways that are described as 'intensive'. One alternative to intensive poultry farming is free-range farming, however, this method of husbandry also uses large flock sizes in high stocking densities. Friction between supporters of these two main methods of poultry farming has led to long-term issues of ethical consumerism.

Opponents of intensive farming argue that it harms the environment and creates health risks, as well as abusing the animals. Advocates of intensive farming say that their highly efficient systems save land and food resources due to increased productivity, stating that the animals are looked after in state-of-the-art environmentally controlled facilities. The most intensive poultry farming methods are very efficient and allow meat and eggs to be available to the consumer in all seasons at a lower cost than free-range production. Poultry producers routinely use nationally approved medications, such as antibiotics, in feed or drinking water, to treat disease or to prevent disease outbreaks. Some FDA-approved medications are also approved for improved feed utilization.

### Egg-laying Chickens - husbandry Systems

Commercial hens usually begin laying eggs at 16–20 weeks of age, although production gradually declines soon after from approximately 25 weeks of age. This means that in many countries, by approximately 72 weeks of age, flocks are considered economically unviable and are slaughtered after approximately 12 months of egg production, although chickens will naturally live for 6 or more years. In some countries, hens are force moulted to re-invigorate egg-laying.

Environmental conditions are often automatically controlled in egg-laying systems. For example, the duration of the light phase is initially increased to prompt the beginning of egg-laying at 16–20 weeks of age and then mimics summer daylength which stimulates the hens to continue laying all year round; normally, egg production occurs only in the warmer months. Some commercial breeds of hen can produce over 300 eggs a year. Critics argue that year-round egg production stresses the birds more than normal seasonal production.

### Free-range Laying Hens

Free-range poultry farming allows the birds to roam freely for a period of the day, although they are usually confined in sheds at night to protect them from predators or kept indoors if the weather is particularly bad. In the UK, the Department for Environment, Food and Rural Affairs (Defra) states that a free-range chicken must have daytime access to open-air runs during at least half of its life.

Unlike in the United States, this definition also applies to egg laying hens. The European Union regulates marketing standards for egg farming which specifies a minimum condition for free-range eggs that "hens have continuous daytime access to open-air runs, except in the case of temporary restrictions imposed by veterinary authorities". The RSPCA "Welfare standards for laying hens and pullets" indicates that the stocking rate must not exceed 1,000 birds per hectare (10 $m^2$ per hen) of range available and a minimum area of overhead shade/shelter of 8 $m^2$ per 1,000 hens must be provided.

Free-range farming of egg-laying hens is increasing its share of the market. Defra figures indicate that 45% of eggs produced in the UK throughout 2010 were free-range, 5% were produced in barn systems and 50% from cages. This compares with 41% being free-range in 2009.

Finding suitable land with adequate drainage to minimise worms and coccidial oocysts, suitable protection from prevailing winds, good ventilation, access and protection from predators can be difficult. Excess heat, cold or damp can have a harmful effect on the animals and their productivity. Unlike cage and barn systems, free-range farmers have little control over the food their animals come across, which can lead to unreliable productivity. In some farms, the manure from free-range poultry can be used to benefit crops.

The benefits of free-range poultry farming for laying hens include opportunities for natural behaviours such as pecking, scratching, foraging and exercise outdoors.

Free-range farming has come under criticism concerning animal welfare. Cannibalism, feather pecking and vent pecking, although not limited to free-range systems, can be common requiring beak trimming as a preventative measure. Diseases can be common and the animals are vulnerable to predators. In South-East Asia, a lack of disease control in free range farming has been associated with outbreaks of Avian influenza.

### Organic Laying Hens

In organic egg-laying systems, chickens are also free-range. Organic systems are based upon restrictions on the routine use of synthetic yolk colourants, in-feed or in-water medications, other food additives and synthetic amino acids, and a lower stocking density and smaller group sizes. The Soil Association standards used to certify organic flocks in the UK, indicate a maximum outdoors stocking density of 1,000 birds per hectare and a maximum of 2,000 hens in each poultry house. In the UK, organic laying hens are not routinely beak-trimmed.

### Yarding for Laying Hens

While often confused with free-range farming, yarding is actually a separate method of poultry culture by which chickens and cows are raised together. The distinction is that free-range poultry are either totally unfenced, or the fence is so distant that it has little influence on their freedom of movement. Yarding is common technique used by small farms in the Northeastern US. The birds are released daily from hutches or coops. The hens usually lay eggs either on the floor of the coop or in baskets if provided by the farmer. This husbandry technique can be complicated if used with roosters, mostly because of aggressive behaviour.

### Battery Cages for Laying Hens

The majority of hens in many countries are reared in battery cages, although the European Union Council Directive 1999/74/EC has banned the conventional battery cage in EU states from January 2012. These are small cages, usually made of metal in modern systems, housing 3 to 8 hens. The walls are made of either solid metal or mesh, and the floor is sloped wire mesh to allow the faeces to drop through and eggs to roll onto an egg-collecting conveyor belt. Water is usually provided by overhead nipple systems, and food in a trough along the front of the cage replenished at regular intervals by a mechanical chain. The cages are arranged in long rows as multiple tiers, often with cages back-to-back (hence the term 'battery cage').

Within a single shed, there may be several floors containg battery cages meaning that a single shed may contain many tens of thousands of hens. Light intensity is often kept low (e.g. 10 lux) to reduce feather pecking and vent pecking. Benefits of battery cages include easier care for the birds, floor eggs which are expensive to collect are eliminated, eggs are cleaner, capture at the end of lay is expedited, generally less feed is required to produce eggs, broodiness is eliminated, more hens may be housed in a given house floor space, internal parasites are more easily treated, and labour requirements are generally much reduced.

In farms using cages for egg production, there are more birds per unit area; this allows for greater productivity and lower food costs. Floor space ranges upwards from 300 $cm^2$ per hen. EU standards in 2003 called for at least 550 $cm^2$ per hen. In the US, the current recommendation by the United Egg Producers is 67 to 86 $in^2$ (430 to 560 $cm^2$) per bird.

The space available to battery hens has often been described as less than the size of a piece of A4 paper. Animal welfare scientists have been critical of battery cages because they do not provide hens with sufficient space to stand, walk, flap their wings, perch, or make a nest, and it is widely considered that hens suffer through boredom and frustration through being unable to perform these behaviours. This can lead to a wide range of abnormal behaviours, some of which are injurious to the hens or their cagemates.

### *Furnished Cages for Laying Hens*

In 1999, the European Union Council Directive 1999/74/EC banned conventional battery cages for laying hens throughout the European Union from January 1, 2012; they were banned previously in other countries including Switzerland. In response to these bans, development of prototype commercial furnished cage systems began in the 1980s. Furnished cages, sometimes called 'enriched' or 'modified' cages, are cages for egg laying hens which have been designed to overcome some of the welfare concerns of battery cages whilst retaining their economic and husbandry advantages, and also provide some of the welfare advantages of non-cage systems.

Many design features of furnished cages have been incorporated because research in animal welfare science has shown them to be of benefit to the hens. In the UK, the Defra "Code for the Welfare of Laying Hens" states furnished cages should provide at least 750 $cm^2$ of cage area per hen, 600 $cm^2$ of which should be usable; the height of the cage other than that above the usable area should be at least 20 cm at every point and no cage should have a total area that is less than 2000 $cm^2$. In addition, furnished cages should provide a nest, litter such that pecking and scratching are possible, appropriate perches allowing at least 15 cm per hen, a claw-shortening device, and a feed trough which may be used without restriction providing 12 cm per hen.

Modern egg laying breeds often suffer from osteoporosis which results in the chicken's skeletal system being weakened. During egg production, large amounts of calcium are transferred from bones to create egg-shell. Although dietary calcium levels are adequate, absorption of dietary calcium is not always sufficient, given the intensity of production, to fully replenish bone calcium. This can lead to increases in bone breakages, particularly when the hens are being removed from cages at the end of laying.

## Meat-producing Chickens - husbandry Systems

### *Indoor Broilers*

Meat chickens, commonly called broilers, are floor-raised on litter such as wood shavings or rice hulls, indoors in climate-controlled housing. Under modern farming methods, meat chickens reared indoors reach slaughter weight at 5 to 6 weeks of age.

Broilers are not raised in cages. They are raised in large, open structures known as growout houses. These houses are equipped with mechanical systems to deliver feed and water to the birds. They have ventilation systems and heaters that function as needed. The floor of the house is covered with bedding material consisting of wood chips, rice hulls, or peanut shells. Because dry bedding helps maintain flock health, most growout houses have enclosed watering systems ("nipple drinkers") which reduce spillage.

Keeping birds inside a house protects them from predators such as hawks and foxes. Some houses are equipped with curtain walls, which can be rolled up in good weather to admit natural light and fresh air. Most growout houses built in recent years feature "tunnel ventilation," in which a bank of fans draws fresh air through the house.

Traditionally, a flock of broilers consist of about 20,000 birds in a growout house that measures 400 feet long and 40 feet wide, thus providing about eight-tenths of a square foot per bird. The Council for Agricultural Science and Technology (CAST) states that the minimum space is one-half square foot per bird. More modern houses are often larger and contain more birds, but the floor space allotment still meets the needs of the birds.

Because broilers are relatively young and have not reached sexual maturity, they exhibit very little aggressive conduct.

Chicken feed consists primarily of corn and soybean meal with the addition of essential vitamins and minerals. No hormones or steroids are allowed in raising chickens.

### Issues with Indoor Husbandry

In intensive broiler sheds, the air can become highly polluted with ammonia from the droppings. This can damage the chickens' eyes and respiratory systems and can cause painful burns on their legs (called hock burns) and feet. Broilers bred for fast growth have a high rate of leg deformities because the large breast muscles causes distortions of the developing legs and pelvis, and the birds cannot support their increased body weight. Because they cannot move easily, the chickens are not able to adjust their environment to avoid heat, cold or dirt as they would in natural conditions. The added weight and overcrowding also puts a strain on their hearts and lungs and Ascites can develop. In the UK, up to 19 million broilers die in their sheds from heart failure each year.

### *Indoor with Higher Welfare*

Chickens are kept indoors but with more space (around 12 to 14 birds per square metre). They have a richer environment for example with natural light or straw bales that encourage foraging and perching. The chickens grow more slowly and live for up to two weeks longer than intensively farmed birds. The benefits of higher welfare indoor systems are the reduced growth rate, less crowding and more opportunities for natural behaviour.

### *Free-range Broilers*

Free-range broilers are reared under similar conditions to free-range egg laying hens. The breeds grow more slowly than those used for indoor rearing and usually reach slaughter weight at approximately 8 weeks of age. In the EU, each chicken must have one square metre of outdoor space. The benefits of free-range poultry farming include opportunities for natural behaviours such as pecking, scratching, foraging and exercise outdoors. Because they grow slower and have opportunities for exercise, free-range broilers often have better leg and heart health.

### *Organic Broilers*

Organic broiler chickens are reared under similar conditions to free-range broilers but with restrictions on the routine use of in-feed or in-water medications, other food additives and synthetic amino acids. The breeds used are slower growing, more traditional breeds and typically reach slaughter weight at around 12 weeks of age. They have a larger space allowance outside (at least 2 square metres and sometimes up to 10 square metres per bird). The Soil Association standards indicate a maximum outdoors stocking density of 2,500 birds per hectare and a maximum of 1,000 broilers per poultry house.

## Issues with Poultry Farming

### *Humane Treatment*

Animal welfare groups have frequently criticized the poultry industry for engaging in practices which they believe to be inhumane. Many animal rights advocates object to killing chickens for food, the "factory farm conditions" under which they are raised, methods of transport, and slaughter. Compassion Over Killing and other groups have repeatedly conducted undercover investigations at chicken farms and slaughterhouses which they allege confirm their claims of cruelty. Conditions in chicken farms may be unsanitary, allowing the proliferation of diseases such as salmonella and E. coli. Chickens may be raised in very low light intensities, sometimes total darkness, to reduce injurious pecking.

Concerns have been raised that companies growing single varieties of birds for eggs or meat are increasing their susceptibility to disease. Rough handling, crowded transport during various weather conditions and the failure of existing stunning systems to render the birds unconscious before slaughter, have also been cited as welfare concerns.

A common practice among hatcheries for egg-laying hens is the culling of newly-hatched male chicks since they don't lay eggs and do not grow fast enough to be profitable for meat.

### Beak Trimming

Laying hens are routinely beak-trimmed at 1 day of age to reduce the damaging effects of aggression, feather pecking and cannibalism. Scientific studies have shown that beak trimming is likely to cause both acute and chronic pain.

The beak is a complex, functional organ with an extensive nervous supply including nociceptors that sense pain and noxious stimuli. These would almost certainly be stimulated during beak trimming, indicating strongly that acute pain would be experienced. Behavioural evidence of pain after beak trimming in layer hen chicks has been based on the observed reduction in pecking behaviour, reduced activity and social behaviour, and increased sleep duration. Severe beak trimming, or beak trimming birds at an older age, may cause chronic pain. Following beak trimming of older or adult hens, the nociceptors in the beak stump show abnormal patterns of neural discharge, which indicate acute pain.

Neuromas, tangled masses of swollen regenerating axon sprouts, are found in the healed stumps of birds beak trimmed at 5 weeks of age or older and in severely beak trimmed birds. Neuromas have been associated with phantom pain in human amputees and have therefore been linked to chronic pain in beak trimmed birds. If beak trimming is severe because of improper procedure or done in older birds, the neuromas will persist which suggests that beak trimmed older birds experience chronic pain, although this has been debated. Beak-trimmed chicks will initially peck less than non-trimmed chickens, which animal behaviouralist Temple Grandin attributes to guarding against pain. The animal rights activist, Peter Singer, claims this procedure is bad because beaks are sensitive, and the usual practice of trimming them without anaesthesia is considered inhumane by some. Some within the chicken industry claim that beak-trimming is not painful whereas others argue that the procedure causes chronic pain and discomfort, and decreases the ability to eat or drink.

### Antibiotics

Antibiotics have been used on poultry in large quantities since the 1940s, when it was found that the byproducts of antibiotic production, fed because the antibiotic-producing mold had a high level of vitamin $B_{12}$ after the antibiotics were removed, produced higher growth than could be accounted for by the vitamin $B_{12}$ alone. Eventually it was discovered that the trace amounts of antibiotics remaining in the byproducts accounted for this growth.

The mechanism is apparently the adjustment of intestinal flora, favouring "good" bacteria while suppressing "bad" bacteria that provoke inflammation of the gut mucosa. So, the goal of antibiotics as a growth promoter is the same as for probiotics. Because the antibiotics used are not absorbed by the gut, they do not put antibiotics into the meat or eggs. Antibiotics are used routinely in poultry for this reason, and also to prevent and treat disease. Many contend that this puts humans at risk as bacterial strains develop stronger and stronger resistances. Critics point out that, after six decades of heavy agricultural use of antibiotics, opponents of antibiotics must still make arguments about theoretical risks, since actual examples are hard to come by. All antibiotic-resistant strains of human diseases whose origin is known originated in hospitals rather than farms.

A proposed bill in the United States Congress would make the use of antibiotics in animal feed legal only for therapeutic (rather than preventative) use, but it has not been passed. However, this may present the risk of slaughtered chickens harboring pathogenic bacteria and passing them on to humans that consume them.

In October 2000, the U.S. Food and Drug Administration (FDA) discovered that two antibiotics were no longer effective in treating diseases found in factory-farmed chickens; one antibiotic was swiftly pulled from the market, but the other, Baytril, was not. Bayer, the company which produced it, contested the claim and as a result, Baytril remained in use until July 2005.

To prevent any residues of antibiotics in chicken meat, any given antibiotics are required to have a "withdrawal" period before they can be slaughtered. Samples of poultry at slaughter are randomly tested by the FSIS, and shows a very low percentage of residue violations

### *Arsenic*

Poultry feed can also include roxarsone or nitarsone, arsenical antimicrobial drugs that also promote growth. Roxarsone was used as a broiler starter by about 70% of the broiler growers between 1995 to 2000. The drugs have generated controversy because it contains arsenic, which is highly toxic to humans. This arsenic could be transmitted through run-off from the poultry yards. A 2004 study by the U.S. magazine Consumer Reports reported "no detectable arsenic in our samples of muscle" but found "A few of our chicken-liver samples has an amount that according to EPA standards could cause neurological problems in a child who ate 2 ounces of cooked liver per week or in an adult who ate 5.5 ounces per week."

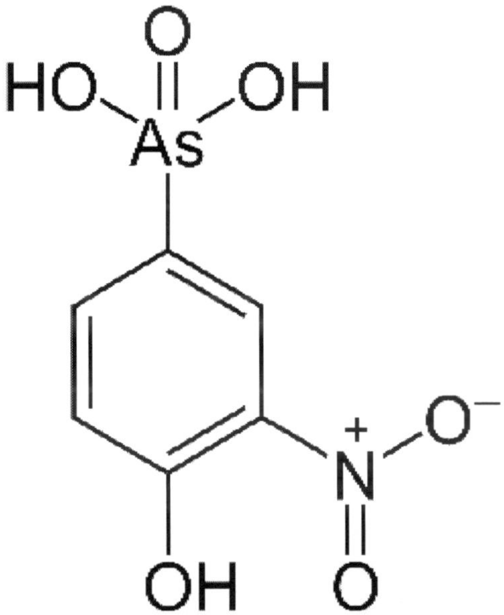


***Figure:*** *Roxarsone, a controversial arsenic compound used as a nutritional supplement for chickens.*

The U.S. Food and Drug Administration (FDA), however, is the organization responsible for the regulation of foods in America, and all samples tested were "far less than the... amount allowed in a food product."

### Growth Hormones

Hormone use in poultry production is illegal in the United States. Similarly, no chicken meat for sale in Australia is fed hormones. Several scientific studies have documented the fact that chickens grow rapidly because they are bred to do so, not because of growth hormones. A small producer of natural and organic chickens confirmed this assumption:

"Using hormones to boost egg production was a brief fad in the Forties, but was abandoned because it didn't work. Using hormones to produce soft-meated roasters lasted into the Fifties, but the improved growth rates of normal, untreated broilers made the practice irrelevant—the broilers got as big as anyone wanted without chemicals. The only hormone that was ever used in any quantity on poultry (DES) was banned in 1959, and everyone but a few die-hard farmers had given up hormones by then, anyway. Hormones are now illegal in poultry and eggs. "

### E. Coli

According to Consumer Reports, "1.1 million or more Americans [are] sickened each year by undercooked, tainted chicken." A USDA study discovered *E. coli (Biotype I)* in 99% of supermarket chicken, the result of chicken butchering not being a sterile process. However, the same study also shows that the strain of *E. coli* found was always a non-lethal form, and no chicken had any of the pathenogenic O157:H7 serotype. Many of these chickens, furthermore, had relatively low levels of contamination.

Feces tend to leak from the carcass until the evisceration stage, and the evisceration stage itself gives an opportunity for the interior of the carcass to receive intestinal bacteria. (So does the skin of the carcass, but the skin presents a better barrier to bacteria and reaches higher temperatures during cooking). Before 1950, this was contained largely by not eviscerating the carcass at the time of butchering, deferring this until the time of retail sale or in the home.

This gave the intestinal bacteria less opportunity to colonize the edible meat. The development of the "ready-to-cook broiler" in the 1950s added convenience while introducing risk, under the assumption that end-to-end refrigeration and thorough cooking would provide adequate protection. *E. coli* can be killed by proper cooking times, but there is still some risk associated with it, and its near-ubiquity in commercially farmed chicken is troubling to some. Irradiation has been proposed as a means of sterilizing chicken meat after butchering.

### *Avian Influenza*

There is also a risk that crowded conditions in chicken farms will allow avian influenza (bird flu) to spread quickly. A United Nations press release states: "Governments, local authorities and international agencies need to take a greatly increased role in combating the role of factory-farming, commerce in live poultry, and wildlife markets which provide ideal conditions for the virus to spread and mutate into a more dangerous form..."

### *Efficiency*

Farming of chickens on an industrial scale relies largely on high protein feeds derived from soyabeans; in the European Union the soybean dominates the protein supply for animal feed, and the poultry industry is the largest consumer of such feed. Two kilograms of grain must be fed to poultry to produce 1 kg of weight gain, much less than that required for pork or beef. However, for every gram of protein consumed, chickens yield only 0.33 g of edible protein.

### *Economic Factors*

Changes in commodity prices for poultry feed have a direct effect on the cost of doing business in the poultry industry. For instance, a significant rise in the price of corn in the United States can put significant economic pressure on large industrial chicken farming operations.

## World Chicken Population

The Food and Agriculture Organization of the United Nations estimated that in 2002 there were nearly sixteen billion chickens in the world, counting a total population of 15,853,900,000. The figures from the *Global Livestock Production and Health Atlas* for 2004 were as follows:

1. China (3,860,000,000)
2. United States (1,970,000,000)
3. Indonesia (1,200,000,000)
4. Brazil (1,100,000,000)
5. India (648,830,000)

6. Mexico (540,000,000)
7. Russia (340,000,000)
8. Japan (286,000,000)
9. Iran (280,000,000)
10. Turkey (250,000,000)
11. Bangladesh (172,630,000)
12. Nigeria (143,500,000)

In 2009 the annual chicken population in factory farms was estimated at 50 billion, with 6 billion raised in the European Union, over 9 billion raised in the United States and more than 7 billion in China.

### Egg (Food)

Eggs are laid by females of many different species, including birds, reptiles, amphibians, and fish, and have been eaten by mankind for thousands of years. Bird and reptile eggs consist of a protective eggshell, albumen (egg white), and vitellus (egg yolk), contained within various thin membranes. Popular choices for egg consumption are chicken, duck, quail, roe, and caviar, but the egg most often consumed by humans is the chicken egg, by a wide margin.

Egg yolks and whole eggs store significant amounts of protein and choline, and are widely used in cookery. Due to their protein content, the United States Department of Agriculture (USDA) categorizes eggs as *Meats* within the Food Guide Pyramid. Despite the nutritional value of eggs, there are some potential health issues arising from egg quality, storage, and individual allergies.

Chickens and other egg-laying creatures are widely kept throughout the world, and mass production of chicken eggs is a global industry. In 2009, an estimated 62.1 million metric tons of eggs were produced worldwide from a total laying flock of approximately 6.4 billion hens. There are issues of regional variation in demand and expectation, as well as current debates concerning methods of mass production, with the European Union's ban on battery farming of chickens.

### History

Bird eggs have been valuable foodstuffs since prehistory, in both hunting societies and more recent cultures where birds were domesticated. The chicken was probably domesticated for its eggs from jungle fowl native to tropical and subtropical Southeast Asia and

India before 7500 BCE. Chickens were brought to Sumer and Egypt by 1500 BCE, and arrived in Greece around 800 BCE, where the quail had been the primary source of eggs. In Thebes, Egypt, the tomb of Haremhab, built about 1420 BCE, shows a depiction of a man carrying bowls of ostrich eggs and other large eggs, presumably those of the pelican, as offerings. In ancient Rome, eggs were preserved using a number of methods, and meals often started with an egg course. The Romans crushed the shells in their plates to prevent evil spirits from hiding there. In the Middle Ages, eggs were forbidden during Lent because of their richness. The word mayonnaise possibly was derived from *moyeu*, the medieval French word for the yolk, meaning centre or hub.

Egg scrambled with acidic fruit juices were popular in France in the 17th century; this may have been the origin of lemon curd.

The dried egg industry developed in the 19th century, before the rise of the frozen egg industry. In 1878, a company in St. Louis, Missouri started to transform egg yolk and white into a light-brown, meal-like substance by using a drying process. The production of dried eggs significantly expanded during World War II, for use by the United States Armed Forces and its allies.

In 1911, the egg carton was invented by Joseph Coyle in Smithers, British Columbia, to solve a dispute about broken eggs between a farmer in Bulkley Valley and the owner of the Aldermere Hotel. Early egg cartons were made of paper.

### Varieties

Bird eggs are a common food and one of the most versatile ingredients used in cooking. They are important in many branches of the modern food industry. The most commonly used bird eggs are those from the chicken. Duck and goose eggs, and smaller eggs, such as quail eggs, are occasionally used as a gourmet ingredient, as are the largest bird eggs, from ostriches. Gull eggs are considered a delicacy in England, as well as in some Scandinavian countries, particularly in Norway. In some African countries, guineafowl eggs are commonly seen in marketplaces, especially in the spring of each year. Pheasant eggs and emu eggs are perfectly edible, but less widely available. Sometimes they are obtainable from farmers, poulterers, or luxury grocery stores. Most wild birds' eggs are protected by laws in many countries, which prohibit collecting or selling them, or permit these only during specific periods of the year.

## Culinary Properties

### *Types of Dishes*

Chicken eggs are widely used in many types of dishes, both sweet and savory, including many baked goods. Some of the most common preparation methods include scrambled, fried, hard-boiled, soft-boiled, omelettes and pickled. They can also be eaten raw, though this is not recommended for people who may be especially susceptible to salmonellosis, such as the elderly, the infirm, or pregnant women. In addition, the protein in raw eggs is only 51% bioavailable, whereas that of a cooked egg is nearer 91% bioavailable, meaning the protein of cooked eggs is nearly twice as absorbable as the protein from raw eggs.

As an ingredient, egg yolks are an important emulsifier in the kitchen, and are also used as a thickener in custards. The albumen, or egg white, contains protein, but little or no fat, and can be used in cooking separately from the yolk. The proteins in egg white allow it to form foams and aerated dishes. Egg whites may be aerated or whipped to a light, fluffy consistency, and are often used in desserts such as meringues and mousse.

Ground egg shells are sometimes used as a food additive to deliver calcium. Every part of an egg is edible, although the eggshell is generally discarded. Some recipes call for immature or unlaid eggs, which are harvested after the hen is slaughtered or cooked while still inside the chicken.

### *Cooking*

Eggs contain multiple proteins which gel at different temperatures within the yolk and the white, and the temperature determines the gelling time. Egg yolk begins to gelify, or solidify, when it reaches temperatures between about 63 and 70 °C (145 and 158 °F). Egg white gels at slightly higher temperatures, about 60 to 80 °C (140 to 176 °F)- the white contains ovalbumin that sets at the highest temperature. However, in practice, in many cooking processes the white gels first because it is exposed to higher temperatures for longer.

Cooking increases the risk of atherosclerosis due to oxidization of the cholesterol contained in the egg yolk.

Salmonella is killed instantly at 160 °F (71 °C), but is also killed from 54.5 °C (130.1 °F) if held there for sufficiently long time periods. To avoid the issue of salmonella, eggs can be pasteurised in shell at

57C for an hour and 15 minutes. Although the white is slightly milkier, the eggs can be used in normal ways. Whipping for meringue takes significantly longer, but the final volume is virtually the same.

If a boiled egg is overcooked, a greenish ring sometimes appears around egg yolk due to the iron and sulphur compounds in the egg. It can also occur with an abundance of iron in the cooking water. The green ring does not affect the egg's taste; overcooking, however, harms the quality of the protein. Chilling the egg for a few minutes in cold water until it is completely cooled may prevent the greenish ring from forming on the surface of the yolk.

### Flavor Variations

Although the age of the egg and the conditions of its storage have a greater influence, the bird's diet does affect the flavor of the egg. For example, when a brown-egg chicken breed eats rapeseed or soy meals, its intestinal microbes metabolize them into fishy-smelling triethylamine, which ends up in the egg. The unpredictable diet of free-range hens will produce unpredictable eggs. Duck eggs tend to have a flavor distinct from, but still resembling, chicken eggs.

Eggs can also be soaked in mixtures to absorb flavor. Tea eggs are steeped in a brew from a mixture of various spices, soy sauce, and black tea leaves to give flavor.

### Storage

Careful storage of edible eggs is extremely important, as an improperly handled egg can contain elevated levels of *Salmonella* bacteria that can cause severe food poisoning. In the US, eggs are washed, and this cleans the shell, but erodes the cuticle. The USDA thus recommends refrigerating eggs to prevent the growth of *Salmonella.*

Refrigeration also preserves the taste and texture. However, uncracked eggs can be left unrefrigerated for several months without spoiling. In Europe, eggs are not usually washed, and the shells are dirtier, however the cuticle is undamaged, and they do not require refrigeration. In the UK in particular, hens are immunised against salmonella, and the eggs are generally safe for 21 days.

### Preservation

The simplest method to preserve an egg is to treat it with salt. Salt draws water out of bacteria and molds, which prevents their growth. The Chinese salted duck egg is made by immersing duck eggs in brine, or coating them individually with a paste of salt and mud or clay. The eggs stop absorbing salt after about a month, having reached osmotic equilibrium. Their yolks take on an orange-red colour and solidify, but the white remains liquid. They are boiled before consumption, and are often served with rice congee.

Another method is to make pickled eggs, by boiling them first and immersing them in a mixture of vinegar, salt, and spices, such as ginger or allspice. Frequently, beetroot juice is added to impart a red colour to the eggs. If the eggs are immersed in it for a few hours, the distinct red, white, and yellow colours can be seen when the eggs are sliced. If marinated for several days or more, the red colour will reach the yolk. If the eggs are marinated in the mixture for several weeks or more, the vinegar will dissolve much of the shell's calcium carbonate and penetrate the egg, making it acidic enough to inhibit the growth of bacteria and molds. Pickled eggs made this way will generally keep for a year or more without refrigeration.

A century egg or hundred-year-old egg is preserved by coating an egg in a mixture of clay, wood ash, salt, lime, and rice straw for several weeks to several months, depending on the method of processing. After the process is completed, the yolk becomes a dark green, cream-like substance with a strong odour of sulphur and ammonia, while the white becomes a dark brown, transparent jelly with a comparatively mild, distinct flavor. The transforming agent in a century egg is its alkaline material, which gradually raises the pH of the egg from around 9 to 12 or more. This chemical process breaks down some of the complex, flavourless proteins and fats of the yolk into simpler, flavourful ones, which in some way may be thought of as an "inorganic" version of fermentation.

### Cooking Substitutes

For those who do not consume eggs, alternatives used in baking include other rising agents or binding materials, such as ground flax seeds or potato starch flour. Tofu can also act as a partial binding agent, since it is high in lecithin due to its soy content. Applesauce can be used, as well as arrowroot and banana. Extracted soybean lecithin, in turn, is often used in packaged foods as an inexpensive substitute for egg-derived lecithin.

Other egg substitutes are made from just the white of the egg for those who worry about the high cholesterol and fat content in eggs. These products usually have added vitamins and minerals, as well as vegetable-based emulsifiers and thickeners such as xanthan gum or guar gum. These allow the product to maintain the nutrition and several culinary properties of real eggs, making possible foods such as Hollandaise sauce, custard, mayonnaise, and most baked goods with these substitutes.

### Nutritional Value

Eggs add protein to a person's diet, as well as various other nutrients.

Chicken eggs are the most commonly eaten eggs. They supply all essential amino acids for humans (a source of 'complete protein'), and provide several vitamins and minerals, including retinol (vitamin A), riboflavin (vitamin $B_2$), folic acid (vitamin $B_9$), vitamin $B_6$, vitamin $B_{12}$, choline, iron, calcium, phosphorus and potassium. Although not as abundant as red meats, eggs are a source of CoQ10 depending on how they are prepared.

All of the egg's vitamins A, D, and E are in the egg yolk. The egg is one of the few foods to naturally contain vitamin D. A large egg yolk contains approximately 60 Calories (250 kilojoules); the egg white contains about 15 Calories (60 kilojoules). A large yolk contains more than two-thirds of the recommended daily intake of 300 mg of cholesterol (although one study indicates the human body may not absorb much cholesterol from eggs). The yolk makes up about 33% of the liquid weight of the egg. It contains all of the fat, slightly less than half of the protein, and most of the other nutrients. It also contains all of the choline, and one yolk contains approximately half of the recommended daily intake. Choline is an important nutrient for development of the brain, and is said to be important for pregnant and nursing women to ensure healthy fetal brain development.

The diet of the laying hens can greatly affect the nutritional quality of the eggs. For instance, chicken eggs that are especially high in omega 3 fatty acids are produced by feeding laying hens a diet containing polyunsaturated fats and kelp meal. Pasture-raised free-range hens which forage largely for their own food also tend to produce eggs with higher nutritional quality in having less cholesterol and fats while being several times higher in vitamins and omega 3 fatty acids than standard factory eggs. Focusing on the protein and crude fat content, a 2010 USDA study determined there were no significant differences of these two macronutrients in consumer chicken eggs.

Cooked eggs are easier to digest, as well as having a lower risk of salmonellosis.

### Anatomy and Characteristics

The shape of an egg resembles a prolate spheroid with one end larger than the other, with cylindrical symmetry along the long axis.

An egg is surrounded by a thin, hard shell. Inside, the egg yolk is suspended in the egg white by one or two spiral bands of tissue called the chalazae (from the Greek word meaning hailstone or hard lump).

### Air Cell

The larger end of the egg contains the air cell that forms when the contents of the egg cool down and contract after it is laid. Chicken eggs are graded according to the size of this air cell, measured during candling. A very fresh egg has a small air cell and receives a grade of AA. As the size of the air cell increases, and the quality of the egg decreases, the grade moves from AA to A to B. This provides a way of testing the age of an egg: as the air cell increases in size, the egg becomes less dense and the larger end of the egg will rise to increasingly shallower depths when the egg is placed in a bowl of water. A very old egg will actually float in the water and should not be eaten.

### Shell

Egg shell colour is caused by pigment deposition during egg formation in the oviduct and can vary according to species and breed, from the more common white or brown to pink or speckled blue-green. In general, chicken breeds with white ear lobes lay white eggs, whereas chickens with red ear lobes lay brown eggs. Although there is no significant link between shell colour and nutritional value, there is often a cultural preference for one colour over another.

### White

White is the common name for the clear liquid (also called the albumen or the glair/glaire) contained within an egg. In chickens it is formed from the layers of secretions of the anterior section of the hen's oviduct during the passage of the egg. It forms around either fertilized or unfertilized yolks. The primary natural purpose of egg white is to protect the yolk and provide additional nutrition for the growth of the embryo. Egg white consists primarily of about 90% water into which is dissolved 10% proteins (including albumins, mucoproteins, and globulins). Unlike the yolk, which is high in lipids (fats), egg white contains almost no fat, and the carbohydrate content is less than 1%. Egg white has many uses in food, and many others, including the preparation of vaccines such as those for influenza.

### *Yolk*

The yolk in a newly laid egg is round and firm. As the yolk ages, it absorbs water from the albumen, which increases its size and causes it to stretch and weaken the vitelline membrane (the clear casing enclosing the yolk). The resulting effect is a flattened and enlarged yolk shape.

Yolk colour is dependent on the diet of the hen; if the diet contains yellow/orange plant pigments known as xanthophylls, then they are deposited in the yolk, colouring it. Lutein is the most abundant pigment in egg yolk. A colourless diet can produce an almost colourless yolk. Yolk colour is, for example, enhanced if the diet includes products such as yellow corn and marigold petals. In the US, the use of artificial colour additives is forbidden.

## Abnormalities

### *Health Studies*

***Cholesterol and Fat:*** More than half the calories found in eggs come from the fat in the yolk; a large (50 gram) chicken egg contains approximately 5 grams of fat. People on a low-cholesterol diet may need to reduce egg consumption; however, only 27% of the fat in egg is saturated fat (palmitic, stearic and myristic acids) that contains LDL cholesterol. The egg white consists primarily of water (87%) and protein (13%) and contains no cholesterol and little, if any, fat.

There is debate over whether egg yolk presents a health risk. Some research suggests dietary cholesterol increases the ratio of total to HDL cholesterol and, therefore, adversely affects the body's cholesterol profile; whereas other studies show that moderate consumption of eggs, up to one a day, does not appear to increase heart disease risk in healthy individuals.

Harold McGee argues that the cholesterol in the yolk is not what causes a problem, because fat (in particular, saturated) is much more likely to raise cholesterol levels than the actual consumption of cholesterol. In a randomized trial, Sacks et al. found that adding one egg per day to the usual diet of 17 lactovegetarians whose habitual cholesterol intake was very low (97 mg/day) significantly increased LDL cholesterol level by 12%.

### Type 2 Diabetes

Studies have shown conflicting results about a possible connection between egg consumption and type two diabetes. A 1999 prospective study of over 117,000 people by the Harvard School of Public Health concluded, in part, that "The apparent increased risk of CHD associated with higher egg consumption among diabetic participants warrants further research." A 2008 study by the Physicians' Health Study I (1982–2007) and the Women's Health Study (1992–2007) determined the "data suggest that high levels of egg consumption (daily) are associated with an increased risk of type 2 diabetes." However, a study published in 2010 found no link between egg consumption and type 2 diabetes.

### Cardiovascular Risk

Eggs are one of the largest sources of phosphatidylcholine (lecithin) in the human diet. A study published in the scientific journal Nature showed that dietary phosphatidylcholine is digested by bacteria in the gut and eventually converted into the compound TMAO, a compound linked with increased heart disease.

The 1999 Harvard School of Public Health study of 37,851 men and 80,082 women concluded that its "findings suggest that consumption of up to 1 egg per day is unlikely to have substantial overall impact on the risk of CHD or stroke among healthy men and women." In a study of 4,000 people published in the New England Journal of Medicine, scientists found that eating eggs lead to significantly increased levels of TMAO in the blood of study participants and that this in turn led to significantly higher risk of heart attack and stroke after three years of follow-up.

A 2007 study of nearly 10,000 adults demonstrated no correlation between moderate (six per week) egg consumption and cardiovascular disease or strokes, except in the subpopulation of diabetic patients who presented an increased risk of coronary heart disease. One potential alternative explanation for the null finding is that background dietary cholesterol may be so high in the usual Western diet that adding somewhat more has little further effect on blood cholesterol. Other research supports the idea that a high egg intake increases cardiovascular risk in diabetic patients. A 2009 prospective cohort study of over 21,000 individuals suggests that "egg consumption up to 6/week has no major effect on the risk of CVD and mortality and that consumption of 7+/week is associated with a modest increased risk of total mortality" in males, whereas among males with diabetes, "any egg consumption is associated with an increased risk of all-cause mortality and there was suggestive evidence for an increased risk of MI and stroke". A meta-analysis published in the British Medical Journal in 2013 found no association between egg consumption and heart disease or stroke.

### Contamination

A health issue associated with eggs is contamination by pathogenic bacteria, such as *Salmonella enteritidis*. Contamination of eggs exiting a female bird via the cloaca may also occur with other members of the *Salmonella* genus, so care must be taken to prevent the egg shell from becoming contaminated with fecal matter. In commercial practice in the US, eggs are quickly washed with a sanitizing solution within minutes of being laid. The risk of infection from raw or undercooked eggs is dependent in part upon the sanitary conditions under which the hens are kept.

Health experts advise people to refrigerate washed eggs, use them within two weeks, cook them thoroughly, and never consume raw eggs. As with meat, containers and surfaces that have been used to process raw eggs should not come in contact with ready-to-eat food.

A study by the U.S. Department of Agriculture in 2002 (Risk Analysis April 2002 22(2):203-18) suggests the problem is not as prevalent as once thought. It showed that of the 69 billion eggs produced annually, only 2.3 million are contaminated with *Salmonella*—equivalent to just one in every 30,000 eggs—thus showing *Salmonella* infection is quite rarely induced by eggs.

However, this has not been the case in other countries, where *Salmonella enteritidis* and *Salmonella typhimurium* infections due to egg consumptions are major concerns. Egg shells act as hermetic seals that guard against bacteria entering, but this seal can be broken through improper handling or if laid by unhealthy chickens. Most forms of contamination enter through such weaknesses in the shell. In the UK, the British Egg Industry Council award the lions stamp to eggs that, among other things, come from hens that have been vaccinated against *Salmonella*.

### Food Allergy

One of the most common food allergies in infants is eggs. Infants usually have the opportunity to grow out of this allergy during childhood, if exposure is minimized. Allergic reactions against egg white are more common than reactions against egg yolks.

In addition to true allergic reactions, some people experience a food intolerance to egg whites. Food labelling practices in most developed countries now include eggs, egg products and the processing of foods on equipment that also process foods containing eggs in a special allergen alert section of the ingredients on the labels.

### Farming Issues

Most commercially farmed chicken eggs intended for human consumption are unfertilized, since the laying hens are kept without roosters. Fertile eggs can be eaten, with little nutritional difference to the unfertilized. Fertile eggs will not contain a developed embryo, as refrigeration temperatures inhibit cellular growth for an extended time. Sometimes an embryo is allowed to develop but eaten before hatching as with balut.

### Grading by Quality and Size

The US Department of Agriculture grades eggs by the interior quality of the egg and the appearance and condition of the egg shell. Eggs of any quality grade may differ in weight (size).

- U.S. Grade AA
    - Eggs have whites that are thick and firm; yolks that are high, round, and practically free from defects; and clean, unbroken shells.
    - Grade AA and Grade A eggs are best for frying and poaching, where appearance is important.
- U.S. Grade A
    - Eggs have characteristics of Grade AA eggs except the whites are "reasonably" firm.
    - This is the quality most often sold in stores.
- U.S. Grade B
    - Eggs have whites that may be thinner and yolks that may be wider and flatter than eggs of higher grades. The shells must be unbroken, but may show slight stains.

- o This quality is seldom found in retail stores because they are usually used to make liquid, frozen, and dried egg products, as well as other egg-containing products.

In Australia and the European Union, eggs are graded by the hen farming method, free range, battery caged, etc.

Chicken eggs are also graded by size for the purpose of sales.

### *Colour of Eggshell*

Although egg colour is a largely cosmetic issue, with no effect on egg quality or taste, it is a major issue in production due to regional and national preferences for specific colours, and the results of such preferences on demand. For example, in most regions of the United States, chicken eggs are generally white. In some parts of the northeast of that country, particularly New England, where a television jingle for years proclaimed "brown eggs are local eggs, and local eggs are fresh!", brown eggs are more common. Local chicken breeds, including the Rhode Island Red, lay brown eggs. Brown eggs are also preferred in Costa Rica, Ireland, France, and the United Kingdom.

In Brazil and Poland, white chicken eggs are generally regarded as industrial, and brown or reddish ones are preferred. Small farms and smallholdings, particularly in economically advanced nations, may sell eggs of widely varying colours and sizes, with combinations of white, brown, speckled (red), green, and blue eggs in the same box or carton, while the supermarkets at the same time sell mostly eggs from the larger producers, of the colour preferred in that nation or region.

These cultural trends have been observed for many years. *The New York Times* reported during the Second World War that housewives in Boston preferred brown eggs and those in New York preferred white eggs. In February 1976, the British New Scientist magazine, in discussing issues of chicken egg colour, stated "Housewives are particularly fussy about the colour of their eggs, preferring even to pay more for brown eggs although white eggs are just as good".

As a result of these trends, brown eggs are usually more expensive to purchase in regions where white eggs are considered 'normal', due to lower production. In France and the United Kingdom it is very difficult to buy white eggs, with most supermarkets supplying only the more popular brown eggs. By direct contrast, in Egypt it is very hard to source brown eggs, as demand is almost entirely for white ones, with the country's largest supplier describing white eggs as "table eggs" and packaging brown eggs for export.

Research conducted in France in the 1970s demonstrated blue chicken eggs (as laid by certain breeds, including araucanas, heritage skyline, and cream legbar) can be stronger and more resilient to breakage, yet an article in New Scientist magazine (contemporary with that research) stated there was little to no demand for blue-coloured eggs from housewives, despite the clear advantages.

Research at Nihon University, Japan in 1990 revealed a number of different issues were important to Japanese housewives when deciding which eggs to buy; however, colour was a distinct factor, with most Japanese housewives preferring the white colour.

Egg producers carefully consider cultural issues, as well as commercial ones, when selecting the breed or breeds of chicken used for production, as egg colour varies between breeds. Among producers and breeders, brown eggs are often referred to as "tinted", while the speckled eggs preferred by some consumers are often referred to as being "red" in colour.

### *Living Conditions of Birds*

Commercial factory farming operations often involve raising the hens in small, crowded cages, preventing the chickens from engaging in natural behaviours, such as wing-flapping, dust-bathing, scratching, pecking, perching and nest-building. Such restrictions can lead to pacing and escape behaviour.

Many hens confined to battery cages, and some raised in cage-free conditions, are debeaked to prevent harming each other and cannibalism. According to critics of the practice, this can cause hens severe pain to the point where some may refuse to eat and starve to death. Some hens may be force molted to increase egg quality and production level after the molting. Molting can be induced by extended feed withdrawal, water withdrawal or controlled lighting programmes.

Laying hens are often slaughtered between 100 and 130 weeks of age, when their egg productivity starts to decline. Due to modern selective breeding, laying hen strains differ from meat production strains. As male birds of the laying strain do not lay eggs and are not suitable for meat production, they are generally killed soon after they hatch. Free-range eggs are considered by some advocates to be an acceptable substitute to factory-farmed eggs. Free-range laying hens are given outdoor access instead of being contained in crowded cages. Questions on the actual living conditions of free-range hens have been raised in the United States of America, as there is no legal definition or regulations for eggs labelled as free-range in that country.

In the United States, increased public concern for animal welfare has pushed various egg producers to promote eggs under a variety of standards. The most widespread standard in use is determined by United Egg Producers through their voluntary programme of certification. The United Egg Producers programme includes guidelines regarding housing, food, water, air, living space, beak trimming, molting, handling, and transportation, however, opponents such as The Humane Society have alleged UEP Certification is misleading and allows a significant amount of unchecked animal cruelty. Other standards include "Cage Free", "Natural", "Certified Humane", and "Certified Organic". Of these standards, "Certified Humane", which carries requirements for stocking density and cage-free keeping and so on, and "Certified Organic", which requires hens to have outdoor access and be fed only organic vegetarian feed and so on, are the most stringent.

Effective 1 January 2012, the European Union banned conventional battery cages for egg-laying hens, as outlined in EU Directive 1999/74/EC. The EU permits the use of "enriched" furnished cages that must meet certain space and amenity requirements. Egg producers in many member states have objected to the new quality standards while in some countries even furnished cages and family cages are subject to be banned as well. The production standard of the eggs is visible on the mandatory egg marking where the EU egg code begins with 3 for caged chicken to 1 for free-range eggs and 0 for organic egg production.

# 4

# Egg Packaging, Transport and Storage

## Packaging of Shell Eggs

Nature has given the egg a natural package - the shell. Despite its relative strength, the egg is an extremely fragile product and even with the best handling methods, serious losses can result from shell damage. Economical marketing generally requires that eggs be protected by the adoption of specialized packaging and handling procedures.

### *Functions of Packaging*

Packaging is an important component in delivering quality eggs to buyers. It embraces both the art and science of preparing products for storage, transport and eventually sale. Packaging protects the eggs from:

- micro-organisms, such as bacteria;
- natural predators;
- loss of moisture;
- tainting;
- temperatures that cause deterioration; and
- possible crushing while being handled, stored or transported.

Proper handling and storage, as seen in the previous chapter, help control moisture loss, but appropriate packaging may also help prevent it. Eggs also need to breathe, hence the packaging material used must allow for the entrance of oxygen. The material used must be clean and odourless so as to prevent possible contamination and tainting. Authentic egg packaging materials can be reused, but careful attention must be paid to possible damage, odours and cleanliness. The packaging must be made to withstand handling, storage and transport methods of the most diverse kind and to protect the eggs against temperatures that cause deterioration and humidity. Finally, consumers like to see what they are buying, especially if it concerns fresh produce. An egg package should be designed so that the customers not only recognize the product as such, but can also see the eggs they are buying.

Many factors must be taken into consideration for packaging eggs. It is important to obtain information regarding the necessary requirements for a particular market, such as:

- quality maintenance;
- storage facilities;
- type of transport;
- distance to be travelled;
- climatic conditions;
- time involved; and
- costs.

## Egg Packages

There are many different types of egg packages, which vary both in design and packaging material used.

*Type 1.* Packing eggs with clean and odourless rice husks, wheat chaff or chopped straw in a firm walled basket or crate greatly decreases the risk of shell damage. An example of this can be seen in the forefront of Photograph 16.

It is also be possible to pack eggs in a simple basket. The basket has no cushioning material such as straw and therefore damage to the eggs may occur more easily. This kind of packaging may be fit for short distance transport.

*Type 2.* A very common form of packaging is the filler tray. The fillers are then placed in boxes or cases. An example can be seen in Photograph 18.

Filler trays are made of wood pulp moulded to accommodate the eggs. They are constructed so that they can be stacked one on top of the other and can also be placed in boxes ready for transport. Filler trays also offer a convenient method for counting the eggs in each box, without having to count every single egg. Usually the standard egg tray carries 36 eggs. Therefore, if a box holds five trays, for example, the box has a total of 180 eggs (36 x 5 = 180).

The cases used may be made of sawn wood; however, they are more commonly made of cardboard. When using cardboard cases, special care must be taken in stacking so that excessive weight is not placed on a case at the bottom of a stack.

Fillers can also be made of plastic as seen in Photograph 19. The advantages of using plastic egg fillers are that they can be reused and are washable. The fillers can be covered with plastic coverings and be used as packages for final sale to the buyer. More importantly, however, plastic transparent fillers allow for the inspection of eggs without handling or touching the eggs.

*Type 3.* Eggs can also be packed in packages that are smaller and specific for retail sale. Each package can hold from two to twelve eggs. These cases can be made of paperboard or moulded wood pulp as seen in Photograph 20, or can be made of plastic as shown in Photograph 21.

It is also possible to pack eggs in small paperboard cases and cover them with plastic film. Egg cases have also been developed from polystyrene. The advantages of using polystyrene are superior cushioning and protection against odours and moisture. The package is also resistant to fungus and mould growth.

The use of small cases is restricted by availability and cost considerations. However, small cases are good for retailers and customers. They are easy for the retailers to handle and customers are able to inspect the eggs.

### Labelling

Labels are a source of important information for the wholesaler, retailer and consumer and not just pieces of paper stuck onto cartons or boxes. The important facts on the label contain information for buyers concerning the eggs, their size and weight and quality/grade description - AA, A or B. Labels may also indicate the producer, when the eggs were laid, how to store them and their expiration date. Persuading the buyer to purchase the product without tasting, smelling or touching is another function of labelling.

***A Sample Label Can be Seen Here Below:*** Labels can be either printed directly on cartons or attached to the cartons. The cost of labelling must be taken into consideration. Simple methods of labelling are available such as stencil or stamp as can be seen below.

### *Costs of Packaging*

When calculating the costs of packaging, expenses must be considered for:

- packaging materials;
- labelling;
- labour;
- additional working capital required;
- changing existing facilities (if applicable); and
- packaging machinery (if applicable).

## Storage of Eggs

The storage of shell eggs during the main laying season, in order to conserve them for consumption when they are scarce, has been practised for many centuries.

For the successful storage of eggs, the following conditions must be met.

- The eggs placed in storage must be clean; they must not be washed or wet.
- Packaging material used should be new, clean and odourless.
- Loss of water due to evaporation should be reduced to a minimum.
- The storage room must be free from tainting products and materials and should be cleaned regularly with odourless detergent sanitizers.
- The storage room must be kept at a constant temperature and humidity must be checked.
- There should be air circulation in the storage room.
- Eggs should be stored so that they are allowed to breathe.
- As far as possible, interior quality should be monitored; there should be a good proportion of thick white, the yolk should stand up well, and the flavour of white and yolk should be good.

If all of the above requirements are to be met, refrigerated storage is necessary.

### *Cold Storage of Eggs*

In the tropics, eggs can deteriorate very quickly unless they are stored at low temperatures. The ideal temperature for storage in such climates is 13°C or lower (usually between 10° and 13° C). Here refrigeration is a necessity for successful commercial storage; however, it may be unavailable or the costs too high.

The most important factors in successful cold storage are as follows.

- The selection and packaging of eggs.
- The equipment and preparation of the cold store.
- Proper temperature, humidity and air circulation.
- Periodic testing for quality.
- The gradual adjustment of eggs to higher temperatures when removed from storage.

*The selection and packaging of eggs for storage.* Eggs for storage must be clean, of good interior quality and have a sound shell. If they are to be stored for more then a month, they should be equivalent to the U.S. grade A. Therefore, it is best to candle all eggs before storage. It may also be advisable to take a sample and to break out these eggs as a further quality check. The period of time between laying and storage should not be more than a few days. The eggs should be kept cool during that time.

Packaging materials used for storage should be new, clean, odourless and free from damage. When packaging material is reused, it is extremely important that it is clean, odourless and free from damage. It is important that the material used allow the eggs to "breathe" and to be free from tainting odours. It should also be sturdy in the event that the cases have to be stockpiled on top of one another.

*The equipment and preparation of the cold store.* The storage room should have a concrete floor that is washable. Walls and ceilings must also be washable. Wooden buildings have been found to be satisfactory, provided they do not impart foreign odours or flavours to the eggs. The room should be scrubbed thoroughly with hot water and soap or an odourless detergent sanitizer before being used. A final rinse with a hypochlorite solution will help greatly in deodorizing the storeroom. A liberal application of freshly slaked lime to unpainted plaster surfaces will also help. The storage room should be aired and dried out thoroughly after cleaning, then closed up and the refrigeration turned on. It is best to allow several days for the temperature and humidity to stabilize before introducing the eggs.

*Proper temperature, humidity and air circulation.* Careful and accurate control of the air condition is essential. A temperature between

- 1.5° and - 0° C is recommended. At a temperature of - 2.5° C eggs freeze. The room should be well constructed and insulated and the refrigeration should be capable of maintaining an adequate uniform temperature in all areas. The cases of eggs should be separated by wood-strips and kept well away from the walls so as not to obstruct air circulation. Aisles left for the convenience of handling specific egg cases also help air circulation. Periodic ventilation of the storage room is advisable to promote air exchange.

The relative humidity should be between 80 and 85 percent at a cold storage temperature of - 1° C. At cold storage temperatures of about 10° C the relative humidity should be between 75 and 80 percent. In such instances, on average, egg weight loss should not exceed 0.5 percent per month. During the early stages of storage when the packaging material is absorbing moisture at a high rate, the floors should be sprinkled with clean water several times a day. If forced-air circulation is feasible, a controlled temperature water-spray air washer may be used. If the humidity becomes excessive, part of the air can be cycled through a unit containing calcium chloride. Where eggs have been oiled less attention can be paid to the humidity level.

*Periodic testing for quality.* Periodic quality checks are essential if the risk of heavy egg losses is to be avoided. Every month or so a sample of eggs should be selected from the various lots and tested. Usually a sample of about 1 percent of all eggs in storage may be sufficient. For example, if 3 000 eggs are kept in storage, 30 eggs sampled from various egg cases will enable a good estimation of the general quality level of the eggs. If there is evidence of excessive deterioration, it is best to dispose of the eggs quickly, after eliminating those that are unfit for consumption.

*The gradual adjustment of eggs to higher temperatures.* Care must be taken in removing eggs from storage to avoid the condensation of moisture on shells. This is minimized by raising the temperature slowly or by moving the eggs through rooms with intermediate temperatures. If condensation occurs, the eggs should be held under conditions that allow the moisture to evaporate within a day or so.

As indicated earlier, eggs should not be stored with products that may taint them. For the long term, eggs are best stored alone, while for the short term they may be kept with dairy products such as milk and mild-flavoured cheese. The average storage life for eggs is between six and seven months.

## Economics of Cold Storage

There is a tendency to underestimate the difficulties involved in providing good cold storage facilities and to recommend their installation without adequate investigation of their cost and potential economic return. The following factors should be considered when contemplating cold storage.

- Refrigeration is a complex and highly technical business.
- Capital investment and operating costs must be estimated.
- Potential available business must be appraised.

The potential available business must be appraised as well as its distribution over the different seasons of the year and the costs involved. Egg storage to even out the availability of supplies is likely to provide business for only a part of the year. The average - not maximum - price difference between the plentiful and scarce seasons must be calculated. If projected returns do not significantly exceed the costs envisaged for storage, there is little incentive for egg traders to make use of storage.

It must be considered that some egg producers, according to their circumstances and possibilities, maintain yearly production through special breeding and feeding programmes and by providing illumination in the hen laying houses. This may even out the rate of egg production throughout the year and hence long-term storage should not be considered.

## Layout of Packaging and Storing Facilities for Shell Eggs

A model layout of a packaging and storing room for shell eggs is seen below.

1. Eggs enter the packing/storage facility
2. Temporary store room
3. Candling room
4. Weighing/cleaning room
5. Packaging area
6. Long-term storage
7. Eggs ready for transport

The layout for the packaging and storing facility is of great importance for efficient and effective management. The various rooms should be kept clean, well ventilated and, where necessary, refrigeration provided. All personnel working in the facility should wear clean outer garments, use caps or head bands and wash their hands when handling eggs and equipment. All equipment used should be clean.

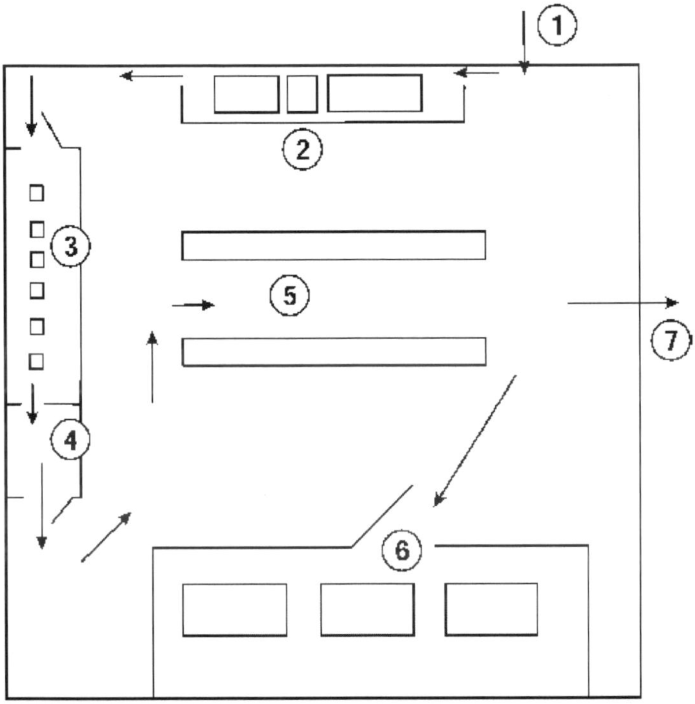


***Figure:*** *Layout of packaging and storage facility*

1. *Eggs enter the packing/storage room.* Eggs from production are brought into the packing/storage facility. Eggs can be brought in by hand or by conveyor belt. In intensive egg production, the birds lay eggs that roll out of the cage onto conveyor belts, which transport the eggs directly to the packing/storage facilities. This can be seen in Photograph 22. Photograph 23 shows how eggs could be stacked for manual movement if brought in by hand from production to the packing/storage facilities.
2. *Temporary storage room.* Here eggs are stored temporarily before they are moved to the candling room.
3. *Candling room.* Eggs are brought into the candling room, where candlers verify the interior and external quality of eggs. The small squares represent candling benches. The candling machine in Photograph 24 is used in fully mechanized or semi-mechanized systems, where the eggs are brought onto the candling machine by a conveyor belt.

4. *Cleaning/weighing room.* After candling, eggs are transported by conveyor belts to the cleaning/weighing room. Here eggs are cleaned with abrasives, where possible, and sorted by weight. Usually the size indicates which category eggs should fall into - small, medium or large. This can be done by hand; however, automated weighing machinery is available.
5. *Packaging area.* After weighing, the eggs are taken to the packaging area. Packaging can be done either by hand or automatically by machinery. As seen in Photograph 26 the eggs arrive at machine No. 1. If we look at Photograph 27, there are three weighing machines numbered one to three. Each machine is set to pack only predetermined egg weights. For example, machine No.1 packs only 60-gram eggs. If the eggs are below that weight, they will be conveyed to machine No. 2. The eggs are then packed automatically.

After packing, eggs may either be kept in long-term storage or may be ready for immediate transport.

### Transport of Eggs

For the successful transport of shell eggs three essential requirements must be met.

1. The containers and packaging materials must be such that the eggs are well protected against mechanical damage.
2. Care should be taken at all stages of handling and transport. Workers handling eggs should be instructed so that they appreciate the need for careful handling. The provision of convenient loading platforms at packing stations, loading depots and railing stations, and handling aids, such as hand trucks and lifts, are of great help.
3. The eggs must be protected at all times against exposure to temperatures that cause deterioration in quality as well as contamination, especially tainting.

The permissible range of temperatures during loading and transport depends on the local climatic conditions and the duration of the journey.

Care is needed to avoid excessive shaking, especially where roads are bad. Egg containers should be stacked tightly and tied down securely to minimize movement. Covers should be used to protect them from the heat of the sun, rain and extreme cold where applicable.

Where bicycles are used, a device such as a special carrier suspended on springs may be helpful.

***Table:*** *Recommended temperatures for loading and transport*

| | ***Transport over 2 or 3 days*** | ***Transport over 5 or 6 days*** |
|---|---|---|
| Maximum on loading | +6° C | +3° C |
| Recommended for transport | -1° to + 3° C | -1° to + 1° C |
| Acceptable for transport | 1° to + 6° C | 1° to + 3° C |

A basic prerequisite for all long-distance transport is that arrangements be made for proper reception, handling and storage at the end of the journey.

This is especially important where large lots are delivered to a relatively small market. Without access to suitable storage facilities, the eggs may have to be marketed quickly under adverse climatic conditions, which may cause substantial quality deterioration and price losses.

Delivery of high quality eggs over long distances, especially in hot climates, generally calls for refrigeration. Requirements for the successful operation of refrigerated transport equipment are rather rigid especially as regards the following factors:

- efficiency and durability of insulation;
- adequacy and reliability of the cooling mechanism; and
- adequate circulation of air within the vehicle or container so that variations of temperature are slight.

Decisions on the establishment of new refrigerated transport services for eggs should be based on thorough economic as well as technical evaluations.

The following criteria should be taken into consideration.

- The need for a managerial and operational staff that is competent in all the operations involved in assembling, loading and distributing.
- The necessity of a sufficient volume of trade throughout the year.
- The possibility of making up loads with other compatible produce, e.g. dairy products.
- The possibility of carrying return loads, once eggs have been distributed.
- The degree to which the demand for refrigerated transport is concentrated geographically.

## Marketing Quality Eggs

### Quality Criteria

Quality determines the acceptability of a product to potential customers. The quality of eggs and their stability during storage are largely determined by their physical structure and chemical composition. It is important therefore that those concerned with the handling of eggs are knowledgeable about this information in order to understand why eggs need to be treated in specific ways and to have a rational basis for day-to-day marketing decisions.

***Composition and Attributes of Eggs:*** An egg consists of shell, membrane, albumen or white and yolk.

*The shell.* The shell of an egg has a rigid yet porous structure. The porous shell has great resistance to the entry of micro-organisms when kept dry and considerable resistance to the loss of moisture by evaporation. The colour of the shell, which may be white or brown depending on the breed of the laying chicken, does not affect quality, flavour, cooking characteristics, nutritional value or shell thickness.

*Shell membrane.* Inside the shell there are two membranes. The outer membrane is attached to the shell, the inner membrane is attached to the albumen or egg white. These two membranes provide a protective barrier against bacterial penetration.

*Air space.* An air space or air cell is a pocket of air usually found at the large end of the egg interior between the outer membrane and the inner membrane. This air cell is created by the contraction of the inner contents while the egg cools and by the evaporation of moisture after the egg has been laid. The air cell increases in size as time passes.

*Egg albumen or white.* The albumen of the egg is composed of the outer thin albumen and the inner firm or thick albumen. The outer thin albumen spreads around the inner firm albumen. The inner firm albumen in high quality eggs stands higher and spreads less than the outer thin albumen.

*White fibrous strips.* These are twisted, cord-like strands of egg white, known as chalazae, which hold the yolk in position. Prominent thick chalazae indicate high quality and freshness.

*Yolk.* The yolk is almost spherical and is surrounded by a colourless membrane. The colour of the yolk varies with the type of feed given to the laying hen. If the laying hen is fed on maize, for example, the yolk will become a bright yellow. The colour of the yolk does not affect the nutritional content.

*Egg weight.* The weight of eggs varies widely depending on many factors such as the breed, the age of the layer and environmental temperature. In Africa, for example, the egg weight may range from 35 to 65 grams, while in Europe it may range from 45 to 70 grams. As a layer gets older the weight of the eggs increase as can be seen.

The components of an egg weighing 60 grams are made up as follows:

- yolk (29%) - 17.4 g
- white (61.5%) - 36.9 g
- shell (9.5%) - 5.6 g

***Nutritional Value:*** Eggs are a good source of high quality protein. They provide important sources of iron, vitamins and phosphorus. As a nutritional source of vitamin D, eggs rank second only to fish liver oils. Eggs are low in calcium, which is discarded in the shell, and contain very little vitamin C.

Eggs provide a unique and well-balanced source of nutrients for persons of all ages. Hard-cooked egg yolks are of great nutritional value as a major supplementary source of iron for infants. When children reach one year of age they may also be given egg whites. Eggs contain significant nutritional value, which is essential during rapid body growth, and therefore are excellent food for young children and teenagers.

Low caloric value, ease of digestibility and high nutrient content make eggs valuable in many therapeutic diets for adults. During convalescence, when bland diets may be required, eggs provide a good nutritious diet. For older people, whose caloric needs are lower, eggs are an easy, inexpensive and nutritious food to prepare and eat.

Availability, modest cost, ease of preparation, popular taste appeal and low caloric value give eggs a primary advantage for human nutritional needs.

### Shell Quality: Texture, Colour, Shape and Condition

The ideal shape of an egg as established by tradition and by practical considerations can be seen in Photograph 6.

Shell quality characteristics that must be considered are as follows:

- cleanliness
- soundness (unbroken)

- smoothness
- shape.

The two most desirable shell qualities, cleanliness and soundness, are largely controlled by the production and handling of eggs. Eggs with shell defects should be removed from eggs destined to the retail trade. Clearly, consumers have adverse reactions to cracked or dirty eggs. Even if the cracks in an egg are only visible when candling, the micro-cracks may have serious consequences on quality. These eggs may be sold locally and possibly only a few hours after lay.

When the membrane is broken as well as the shell, the contents of the eggs can leak, and therefore the only practicable market outlet is sale as egg pulp. If the eggs are dirty, for example, with blood or faeces, consumers will react unfavourably to them. Although shell colour is no indication of quality, consumers in some markets may prefer white eggs or brown eggs. In such circumstances, it is advisable to sort eggs by shell colour.

### *Yolk and Albumen Quality*

In quality eggs the yolk should be round, firm and stand up well, and be yellow in colour.

There is often prejudice against very pale or deeply coloured yolks, however, there are some exceptions. In some Italian markets, for example, red yolks are a strong selling point. The yolk should have a pleasant, mild egg odour and flavour and should be surrounded by a large amount of upstanding thick white with only a small amount of thin white. The egg white should have the normal slightly green-yellow colour, though it may be slightly cloudy in appearance.

Consumers are generally very critical of any abnormal conditions in the egg yolk and white. Factors that may cause loss of quality are as follows:

- natural factors
- temperature
- humidity
- time
- handling
- storage
- tainting.

Natural factors, for example, can be blood spots, which may range from small specks to a square centimetre in size. They may vary in colour from light grey to bright red and may be found in the yolk or in the egg white. "Blood eggs", with blood diffused throughout the white or spread around the yolk, are not commonly found and are generally rejected by the consumer. Photograph 7 shows the various degrees of spotting and blood diffusion.

### Deterioration

The changes that occur in eggs stored for a week to ten days at a temperature between 27° and 29° C are comparable to those that occur in similar eggs in cold storage for several months at a temperature of - 1° C. The effect of temperature and storage on eggs can be seen in Photograph 8. The typical appearance of eggs stored up to 13 weeks at temperatures varying from 10° to 46° C can also be seen in this photograph. In advanced stages of deterioration, the thick white may disappear entirely and the yolk may enlarge to the point where its membranes are so weakened that it breaks when the egg is opened. Changes in odour and flavour take three or four weeks at a temperature of 21° C, or six to seven weeks at a temperature of 10° C to become noticeable to the ordinary consumer.

Temperature, humidity, air movement and storage time can all have adverse effects on interior quality. These factors, if not controlled, can cause loss of moisture in eggs. Loss of water through the porous shell will mean loss of weight. A loss of weight of two to three percent is common in marketing eggs and is hardly noticeable to consumers. However, enlarged air cells and a decreased size of egg contents become noticeable when losses exceed this extent.

Coating eggs with oil and other substances and storing them at low temperatures and high humidity may control moisture loss. The best conditions for storage are at a temperature of about - 1° C and relative humidity between 80 and 85 percent. At a temperature of 10° C, lower relative humidity is needed, between 75 and 80 percent. At all temperatures there is the risk of mould spoilage where the relative humidity is too high. Packaging materials that are too dry or are excessively moist and absorbent will also accentuate evaporation losses.

The contents of eggs when just laid are usually sterile and contain few organisms capable of causing spoilage even when the shells are slightly dirty or stained. The main cause of spoilage by bacteria is the washing of dirty eggs before marketing. When the egg is washed, organisms from water - usually bacteria - can penetrate the shell. Once inside they multiply and eventually spoil the egg, causing green, black and red rots. Even when eggs become wet without any cleaning process, for example, by condensation after removal from refrigerated storage into a warm temperature, conditions may be favourable for the penetration of micro-organisms and rotting may follow. When eggs are kept dry, no such way is provided for bacteria to penetrate the shell.

Mould spores normally present on eggshells may, if sufficient time elapses, germinate and grow, penetrating the shell and causing spoilage. Generally this occurs only when eggs are in cold storage for several months or more under conditions of high humidity (above 85 percent). It can occur, however, at any temperature if the humidity is sufficiently high and the holding time long enough.

Eggs can easily be tainted by strong odours from kerosene, gasoline, diesel oil, paint and varnish, and by such fruit and vegetables as apples, onions and potatoes. Special care must therefore be taken in storage, packaging materials and transport facilities used.

### *Quality Maintenance*

Maintaining fresh egg quality from producer to consumer is one of the major problems facing those engaged in marketing eggs. Proper attention to production, distribution and point-of-sale phases are of vital importance in maintaining egg quality.

### *Production Factors*

The factors that affect egg production are discussed. The main production factors that affect quality maintenance are the following:

- breed
- age
- feed
- management
- disease control
- handling/collecting eggs
- housing.

*Breed.* The breed of the laying hen affects shell colour; for example, Leghorns produce white eggs, while Rhode Island Reds produce brown eggs. The following egg quality factors are partly inherited: shell texture and thickness, the incidence of blood spots and the upstanding quality and relative amount of thick albumen. Though it may not always be possible, a consistent policy of selection for breeds by egg producers can bring noticeable improvements to quality.

*Age.* Birds typically begin producing eggs in their twentieth or twenty-first week and continue for slightly over a year. This is the best laying period and eggs tend to increase in size until the end of the egg production cycle. Birds lay fewer eggs as they near the moulting period. In the second year of lay, eggs tend to be of lower quality.

*Feed.* Egg quality and composition derive primarily from what a layer is fed. In terms of taste, for example, eggs laid by hens fed on fishmeal will have a "fishy" taste. The type of feed will also influence the shell of an egg and the colour of the yolk. Layers must be kept away from certain plant foods if egg colour defects are to be avoided. These may include cottonseed meal and the foliage of the *sterculiaceae* and *malvaceae* such as mallow weed.

Regular access to fresh or high-quality dehydrated green feed helps birds to produce eggs with a uniform yellow yolk. Yellow maize, alfalfa meal, and fresh grass provide good pigment sources for a normal yellowish-orange yolk colour.

*Management.* Good general management of the laying flock can improve egg quality. If birds are treated correctly and not put under conditions of stress they will produce properly.

*Disease control.* Diseases have an effect on egg quality. Infectious bronchitis and Newcastle disease, for example, will cause birds to lay eggs with poor quality shells and with extremely poor quality albumen. Many of the birds continue to lay poor quality eggs even after recovery. Effective vaccines should be administered.

*Handling/collecting eggs.* Frequent collection is essential each day in order to limit the number of dirty and damaged eggs and also to prevent the hens from eating the eggs. Careful handling is necessary in order to avoid breakage.

*Laying house.* The number of dirty eggs produced can be reduced significantly by providing good housing and clean nests for the layers. Cleaning and hygiene operations should be carried out frequently.

## Measures to Prevent Deterioration During Marketing

*Temperature.* By far the most effective method of minimizing deterioration of quality in eggs is to keep them at temperatures below

13° C. Eggs should never be left standing in the sun or in a room that gets very hot at some point in the day, but should be moved into shaded, well-ventilated rooms and underground cellars as quickly as possible.

Various methods to prevent deterioration by temperature are shown below.

1. A simple method is covering eggs with green leaves, so as to reduce temperature.
2. A method commonly used is that of putting eggs in a porous pot where the outside of the pot is kept damp. Great care should be taken, however, to avoid the excess use of water, which could trickle down to the bottom of the pot damaging the eggs at the bottom.
3. Eggs can be kept in a wide-mouthed earthen pot that is buried in the ground up to half of its height. The inside of the pot is lined with a thin layer of grass to prevent the eggs being spoiled by excess moisture. Eggs are placed in the pot as soon as they are collected and the top covered with a thin cloth to facilitate the exchange of air. A layer of sand and earth is spread around the earthen pot and water is sprinkled on it frequently during the day. The eggs are turned once a day to prevent the internal yolk of the egg from sticking to one side of the eggshell. Such a system may reduce the egg temperature by 8° C below the temperature outside the pot.
4. Another method that can be used which is ideal for dry climates makes use of the cooling effect of evaporation. Baskets of eggs are stored in a small wooden or wire-frame cupboard. A water tray is kept on top of the box and pieces of sacking are placed in the tray and arranged so that they hang on all sides of the box. More elaborate versions with arrangements for a steady dripping of water on to the sacking can be developed. In humid areas such devices would be less useful. The maintenance of egg quality in wet tropical areas is extremely difficult without refrigeration.
5. Refrigerated storerooms can be used if electricity is available. An example of a refrigerated storeroom is shown below. If refrigerated storehouses are not economically viable, the use of electric fans may be appropriate.

### Refrigerated Storeroom

The cooler should be shaded from direct sunlight. It is most efficient when air circulates *freely* round it and does not give good results in a closed room. The cooling effect can be intensified by blowing air through damp sacking with an electric fan.

Producers, wholesalers and retailers should move eggs to consumers as quickly as possible to minimize the risk of spoilage. The importance of avoiding delays at all stages in the distribution channel cannot be overemphasized and should be the primary consideration determining marketing arrangements.

*Treatment of dirty eggs.* Some eggs will inevitably have dirty shells. For the purpose of appearance, washing is the most effective and simplest method of removing dirt and stains from the shell surface. The water, however, may contain bacteria that could penetrate the porous eggshell causing it to decay. Odourless detergent-sanitizing substances should be used in the water to wash eggs, but these may be difficult to obtain.

Eggs can be submerged in clean hot water (water temperature should be around 38° C); however, this may cause thermal cracks in the eggshell and internal expansion of the egg content. It is better to avoid washing eggs altogether. Using dry abrasives for scraping and brushing may be the optimal solution. In using this method, care should be taken to avoid removing excessive shell material, which will weaken the shell and increase the rate of evaporation.

The cooler should be shaded from direct sunlight. It is most efficient when air circulates *freely* round it and does not give good results in a closed room. The cooling effect can be intensified by blowing air through damp sacking with an electric fan.

*Shell oiling.* Coating eggs with a thin film of oil greatly reduces losses by evaporation, especially where eggs are in cold storage for several months or are held at temperatures above 21° C. Special odourless, colourless, low-viscosity mineral oils should be used. Where eggs must withstand high temperatures, they should be oiled from four to six hours after lay. If eggs are to be stored at a temperature of 0° C, they should be oiled 18 to 24 hours after lay. Eggs can be oiled by hand dipping wire baskets or by machine. The temperature of the oil should be at least 11° C above that of the eggs. Before the oil is reused it should be heated to a temperature of 116° C to prevent bacteria survival and then be filtered. The oil reservoirs should be cleaned properly. In terms of appearance oiled eggs differ from other eggs only in the slight shine left on the eggshells by the more viscous oils.

## Inducements for Quality Maintenance

Provision of effective incentives for the adoption of quality maintenance procedures is the function of the marketing system. It must provide some means whereby egg quality can be appraised and a system of purchasing premiums and deductions applied accordingly. Methods of assessing quality of eggs are discussed below.

### *Grading and Standardization*

Grading and standardization consist of arranging produce into a number of uniform categories according to physical and quality characteristics of economic importance. It is a process of identification, classification and separation.

The advantages of grading and standardization are as follows.

- Different grade eggs may be sold to different customers. Customers willing to pay more for high quality eggs will be served. On the other hand, eggs with micro-cracks or small blood spots may be sold to bakeries.
- Setting and maintaining a reliable standard creates consumer confidence in the product and a favourable reputation. This will enable buyers (wholesalers, retailers, exporters, consumers) to purchase a reliable product that they recognize and may well avoid inspection and disputes.
- The ability to furnish an accurate description of eggs in storage may help in obtaining credit.

### *Grade Specifications*

The value factors most generally appreciated in eggs are internal quality, appearance and soundness of shell, size and colour.

Most egg marketing systems find it advantageous to adopt grading practices that:

- eliminate inedible and defective eggs;
- separate eggs into high and lower acceptable categories; and
- establish uniform weight classifications.

For example, the grading system used in the United States of America, as recommended by the United States Department of Agriculture (USDA), is summarized.

***Table:*** *Summary of United States grade standards for individual shell eggs*

| *Specifications for each quality factor* | | | |
|---|---|---|---|
| ***Quality Factor*** | ***AA Quality*** | ***A Quality*** | ***B Quality*** |
| Shell | Clean. | Clean. | Clean to slightly stained.* |
| | Unbroken. | Unbroken. | Unbroken. |
| | Practically normal. | Practically normal. | Abnormal. |
| Air Cell | 1/8 inch or less in depth. | 3/16 inch or less in depth. | over 3/16 inch in depth. |
| | Unlimited movement and free or bubbly. | Unlimited movement and free or bubbly. | Unlimited movement and free or bubbly. |
| White | Clear. | Clear. | Weak and watery. |
| | Firm. | Reasonably firm. | Small blood and meat spots present.** |
| Yolk | Outline slightly defined. | Outline fairly well defined. | Outline plainly visible. |
| | | | Enlarged and flattened. |
| | Practically free from defects. | Practically free from defects. | Clearly visible germ development but not blood. |
| | | | Other serious defects. |
| ***For eggs with dirty or broken shells, the standards of quality provide two additional qualities.*** | | | |
| ***Dirty*** | | ***Check*** | |
| Unbroken. Adhering dirt or foreign material, prominent stains, moderately stained areas in excess of B quality. | | Broken or cracked shell, but membranes intact, not leaking.*** | |

* Moderately stained areas permitted (1/32 of surface if localized, or 1/16 if scattered).

** If they are small (aggregating not more than 1/8 inch in diametre).

*** Leaker has broken or cracked shell membranes, and contents leaking or free to leak.

***Note:*** Measurements in inches: 1/8 inch = 3 mm, 3/16 inch = 6 mm.

***Source:*** United States Department of Agriculture

The various egg sizes according to weight used in the United States are as follows:

- Jumbo = 70 g and above
- Extra large = 65-70 g
- Large = 56-65 g
- Medium = 49-56 g
- Small = 42-49 g
- Peewee = 35-42 g

Of course different size specifications and quality factors vary from country to country; for example, the various sizes according to weight used in Africa are as follows:

- Large = 65 g and above
- Medium = 55-65 g
- Small = 45-55 g

## Quality Specifications Development

A simple set of quality specifications might be set up as follows.

*First Grade.* The shell must be clean, unbroken and practically normal in shape and texture. The air cell must not exceed 9.5 mm in depth and may move freely, but not be broken and bubbly. The yolk may appear off-centre, but only slightly enlarged, and may show only slight embryonic development. No foreign objects may be present.

*Second Grade.* The shell must be unbroken, but may be somewhat abnormal in shape and texture. Only slight stains and marks are permitted. The yolk may appear dark and enlarged and may show embryonic development, but not at the blood vessel stage and beyond. Blood spots less then 6mm in diametre are permitted.

*Third grade.* Other edible eggs are permitted, that is, those that are not rotted, mouldy or musty. Also, those eggs that are not incubated to blood vessel stage, and those not containing insects, worms or blood spots 6mm in diametre, are permitted.

## Interior Quality

The most accurate test of interior quality is the break-out method - open the egg on to a flat glass surface and compare the appearance of the yolk and white with that shown for various quality levels as shown.

A more simple side-view comparison of egg quality.

In marketing, however, the above method can only be used on a sample basis. A method that does not require egg breaking is more appropriate.

## Candling

Candling is the only method of testing eggs for quality, internally and externally, without breaking them. It consists of inspecting an egg with a beam of light that makes the interior quality visible. A very simple form of candling is placing a candle in a dark room and positioning an egg in front of the flame and looking at the interior quality. A simple candling device can be seen below. If electricity is available, a light bulb can be placed in the box, otherwise a paraffin lamp or candle can be enclosed in a casing. The hole in the box should be about 3 cm in diametre, sufficient for egg sizes ranging from 40 to 70 grams. The light will shine out of the hole making the interior egg quality visible.

*Candling method.* Pick four eggs to be candled and put two in each hand. Place the first egg near the candling box hole with the large end of the egg held against the light, and with the axis at a 45° angle so that the egg has light shining through it. Twirl the egg so as to observe defects which otherwise might not be observed.

If first egg candled is free of defects, roll the first egg back in to the palm of the hand. Meanwhile, the second egg in the other hand should be brought to the light and examined. While the second egg is up against the light, the third egg in the other hand should be brought into candling position. Photograph 9 shows how to hold eggs while candling.

Although the beginner will soon learn to detect such things as cracked shells and bloodspots, considerable training is needed before internal quality can be estimated with reasonable accuracy. Photograph 10 shows the appearance of eggs of various quality in candling.

The main interior quality points to be observed in candling can be summarized as follows:

*Yolk.* The judgement of internal quality is based mainly on the visibility, ease of movement and shape of yolk. Common yolk faults are the following:

- *Sided* - displaced to an appreciable extent from its normal central position.
- *Stuck* - on twirling the egg, it may be found that the yolk is stuck to the inner shell membrane.
- *Patchy* - uneven in colour, including defects sometimes described as "heat spots."
- *Abnormal in shape* - flattened or irregular and in extreme cases may be broken and dispersed in the white.
- *Discoloured* - of a dark or greyish appearance often with a very distinct outline.
- *Embryonic development* - first shows as a dark halo round the germ cell near the centre of the yolk and later as thin blood vessels and a bright blood ring.

*White.* In practice, the quality of egg white is judged by the degree of movement of the yolk and by the definition of its outline. Common faults in egg white are as follows:

- *Discoloured* - definitely tinted grey, yellow, green or brown.
- *Cloudy* - muddy or streaky. Usually this condition indicates potential rot, but washing an egg in very hot water can cause a similar appearance.

*Air cell.* The depth of the air cell is a rough indication of the age of the egg and there is often a relation between this depth and the internal quality. Hence, the depth of the air cell is taken into account in candling, but other indications of quality are given equal weight. The air cell may be:

- *Large* - exceeding 6 mm in depth.
- *Running* - if the air cell is broken, one or more air bubbles will be found in the white. If the air cell has forced its way between the two shell membranes, bubbles will move around the shell when the egg is candled. A running air cell, however, may be caused by rough handling and should not exclude the egg from a high-grade class.
- *Ringed* - the air cell is very large, sharply defined and with grey or brown edges.

Other common defects of internal quality that may be found are listed here:

- *Blood spots* - clots or streaks of blood in the white or adhering to the yolk.
- *Blood egg* - blood is diffused throughout the white or spread around the yolk.
- *Meat spots* - fatty material, fleshy or liver-like that may be found floating freely in the white, embedded in the chalazae or attached to the yolk.
- *Staleness* - in most cases the air cell is abnormally large, clearly defined and often ringed. As a rule the yolk is sided and its outline clearly defined.
- *Mould growth* - usually grey or black in colour, but can occasionally be pinkish, found on the outside and inside of the shell or shell membranes.
- *Rot* - usually violet, green, red or blue in colour. The early stages of a rot are less easy to detect, but any egg with a

streaky, turbid white should be rejected. The egg may have an unpleasant smell even if unbroken.

- *Taint* - the egg has an abnormal odour.

Photographs 11 through 14 show examples of various egg quality deficiencies that can be seen when candling.

*Shell condition.* Weak, rough, mouldy, cracked and deformed shells may be detected as eggs are picked up for candling. But with candling small or micro-cracks on the eggshell can be seen. A typical deformed eggshell can be seen in Photograph 15.

Another method of verifying shell soundness is that of gently hitting two eggs together (belling). A dull sound instead of clear clinking indicates a cracked egg.

### Developing a Standardized System

Any system used to grade the quality and weight of eggs is only effective if it ensures that consumers obtain eggs of the quality and kind they want. This requires that three conditions be met.

1. The initial classification must be correct.
2. There should be no appreciable deterioration between time of grading and time of sale.
3. The consumers should have a clear guide to the quality of produce they are purchasing.

In developing a grading system, the following preparations are vital.

- Study thoroughly the pattern of production, consumption and trade.
- Work out grade specifications in close consultation with those traders who would be likely to take advantage of them.
- Qualified inspectors are needed to ensure that conformity with the grades indicated is upheld. Producers and packers who accept their inspection can be authorized to apply approved grade stamps.
- Legislation should be enacted to prevent the possibility of misleading labelling. Certain standards may be made obligatory, for example, minimum standards to protect consumers from unwholesome and dangerous foods.
- Finally, if consumers are to take full advantage of a grading system, the grade indications should be clear and easily understood.

It is important that eggs are not allowed to deteriorate below the grade indicated before they reach the consumer. Under conditions of high humidity with temperatures of 32° C and above, eggs being distributed may undergo considerable deterioration in only a few days. Under such circumstances, the inspection of eggs held for more than three or four days must be clearly and responsibly assigned.

### Equipment and Candling Layout

In order to grade and pack eggs with consistent accuracy, speed and economy, it is essential that adequate facilities be provided. These include a semi-darkened room, without stray illumination, that is well equipped with bench and shelf space and with good facilities for handling of eggs and recording results.

The candling device, if possible, should be adjustable and mounted at such a height that the beam of light when emerging horizontally from the device arrives at about the height of the operator's elbow. The lamp or light bulb in the candling box should be kept clean. An operator should use the same candling apparatus every day to help minimize errors. Each operator must have enough room to move freely and handle packing materials and boxes. Efficiency can be increased by installing partitions between candling benches, arranging supplies of packing materials conveniently and by providing for the easy disposal of damaged materials. Floors and walls should have smooth, hard non-reflecting surfaces and be rounded at intersections. Thorough cleanliness is essential to prevent bad odours. Good ventilation should be provided.

The candling bench should be designed to particular needs. Where there are only a few candling benches, operators themselves can obtain eggs and supplies from nearby stocks and carry away completed egg cases. As the volume handled increases, however, it usually becomes more efficient to assign a special worker to service the candlers.

## Chicken (Food)

Chicken is the most common type of poultry in the world, and is prepared as food in a wide variety of ways, varying by region and culture.

### History

The modern chicken is a descendant of Red Junglefowl hybrids along with the Grey Junglefowl first raised thousands of years ago in the northern parts of the Indian subcontinent.

Chicken as a meat has been depicted in Babylonian carvings from around 600 BC. Chicken was one of the most common meats available in the Middle Ages. It was widely believed to be easily digested and considered to be one of the most neutral foodstuff. It was eaten over most of the Eastern hemisphere and a number of different kinds of chicken such as capons, pullets and hens were eaten. It was one of the basic ingredients in the so-called white dish, a stew usually consisting of chicken and fried onions cooked in milk and seasoned with spices and sugar.

Chicken consumption in the United States increased during World War II due to a shortage of beef and pork. In Europe, consumption of chicken overtook that of beef and veal in 1996, linked to consumer awareness of Bovine spongiform encephalopathy (mad cow disease).

## Breeding

Modern varieties of chicken such as the Cornish Cross, are bred specifically for meat production, with an emphasis placed on the ratio of feed to meat produced by the animal. The most common breeds of chicken consumed in the US are Cornish and White Rock.

Chickens raised specifically for food are called broilers. In the United States, broilers are typically butchered at a young age. Modern Cornish Cross hybrids, for example, are butchered as early as 8 weeks for fryers and 12 weeks for roasting birds.Capons (castrated cocks) produce more and fattier meat. For this reason, they are considered a delicacy and were particularly popular in the Middle Ages.

## Edible Components

***Main:***

- Breast: These are white meat and are relatively dry.
- Leg: Comprises two segments:
  1. The "drumstick"; this is dark meat and is the lower part of the leg,
  2. the "thigh"; also dark meat, this is the upper part of the leg.
- Wing: Often served as a light meal or bar food. Buffalo wings are a typical example. Comprises three segments:
  1. the "drumette", shaped like a small drumstick,
  2. the middle "flat" segment, containing two bones, and
  3. the tip, sometimes discarded.

***Figure:*** *Chicken in a public market*

***Other:***

- Chicken feet: These contain relatively little meat, and are eaten mainly for the skin and cartilage.

  Although considered exotic in Western cuisine, the feet are common fare in other cuisines, especially in the Caribbean and China.
- Giblets: organs such as the heart, gizzards, and liver may be included inside a butchered chicken or sold separately.
- Head: Considered a delicacy in China, the head is split down the middle, and the brains and other tissue is eaten.
- Kidneys: Normally left in when a broiler carcass is processed, they are found in deep pockets on each side of the vertebral column.
- Neck: This is served in various Asian dishes.
- Oysters: Located on the back, near the thigh, these small, round pieces of dark meat are often considered to be a delicacy.
- Pygostyle (chicken's buttocks) and testicles: These are commonly eaten in East Asia and some parts of South East Asia.

***By-products:***

- Blood: Immediately after slaughter, blood may be drained into a receptacle, which is then used in various products. In many Asian countries, the blood is poured into low, cylindrical forms, and left to congeal into into disc-like cakes for sale. These are commonly cut into cubes, and used in soup dishes.
- Carcase: After the removal of the flesh, this is used for soup stock.
- Chicken eggs
- Heart and gizzard
- Liver: This is the largest organ of the chicken, and is used in such dishes as Pâté and chopped liver.
- Schmaltz: This is produced by rendering the fat, and is used in various dishes.

## Health

Chicken meat contains about two to three times as much polyunsaturated fat than most types of red meat when measured as weight percentage. Chicken generally includes low fat in the meat itself (castrated roosters excluded). The fat is highly concentrated on the skin. A 100g serving of baked chicken breast contains 4 grams of fat and 31 grams of protein, compared to 10 grams of fat and 27 grams of protein for the same portion of broiled, lean skirt steak.

However according to a 2006 Harvard School of Public Health study of 135,000 people, people who ate grilled skinless chicken 5 or more times a week had a 52 percent higher chance of developing bladder cancer compared to people who did not. However, such strong associations were not found in individuals regularly consuming chicken with skin intact.

## Use of Roxarsone in Chicken Production

In many factory farms, chickens are routinely administered with the feed additive Roxarsone, a relatively benign organoarsenic compound which partially decomposes into inorganic arsenic compounds in the flesh of chickens, and in their feces, which are often used as a fertilizer. The compound is used to control stomach pathogens and promote growth. A *Consumer Reports* study in 2004 reported finding "no detectable arsenic in our samples of muscle" but found "A few of our chicken-liver samples has an amount that according to EPA standards could cause neurological problems in a child who ate 2 ounces of cooked liver per week or in an adult who ate 5.5 ounces per week." However, the amounts found in these livers averaged to 460 part per billion; an amount still less than the 2,000 parts per billion limit set by the FDA. The FDA has found there to be no significant impact for the use of Roxarsone in the production of chicken, turkey or swine.

### Antibiotic Resistance

Information obtained by the Canadian Integrated Programme for Antimicrobial Resistance (CIPARS) "strongly indicates that cephalosporin resistance in humans is moving in lockstep with use of the drug in poultry production." According to the Canadian Medical Association Journal, the unapproved antibiotic ceftiofur is routinely injected into eggs in Quebec and Ontario to discourage infection of hatchlings. Although the data are contested by the industry, antibiotic resistance in humans appears to be directly related to the antibiotic's use in eggs.

A recent study by the Translational Genomics Research Institute showed that nearly half (47%) of the meat and poultry in US grocery stores was contaminated with *S. aureus*, with more than half (52%) of those bacteria resistant to antibiotics.

### Marketing and Sales

Juvenile chickens, of less than 28 days of age at slaughter in the United Kingdom are marketed as poussin. Mature chicken is sold as small, medium or large.

Whole mature chickens are marketed in the United States as fryers, broilers, and roasters. Fryers are the smallest size (2.5-4 lbs dressed for sale), and the most common, as chicken reach this size quickly (about 7 weeks). Most dismembered packaged chicken would be sold whole as fryers. Broilers are larger than fryers. They are typically sold whole. Roasters, or roasting hens, are the largest chickens commonly sold (3–5 months and 6-8 lbs) and are typically more expensive. Even larger and older chickens are called stewing chickens but these are no longer usually found commercially. The names reflect the most appropriate cooking method for the surface area to volume ratio. As the size increases, the volume (which determines how much heat must enter the bird for it to be cooked) increases faster than the surface area (which determines how fast heat can enter the bird). For a fast method of cooking, such as frying, a small bird is appropriate:

frying a large piece of chicken results in the inside being undercooked when the outside is ready.

Chicken is also sold in dismembered pieces. Pieces may include quarters, or fourths of the chicken. A chicken is typically cut into two leg quarters and two breast quarters. Each quarter contains two of the commonly available pieces of chicken. A leg quarter contains the thigh, drumstick and a portion of the back; a leg has the back portion removed. A breast quarter contains the breast, wing and portion of the back; a breast has the back portion and wing removed. Pieces may be sold in packages of all of the same pieces, or in combination packages. Whole chicken cut up refers to either the entire bird cut into 8 individual pieces. (8-piece cut); or sometimes without the back. A 9-piece cut (usually for fast food restaurants) has the tip of the breast cut off before splitting. Pick of the Chicken, or similar titles, refers to a package with only some of the chicken pieces. Typically the breasts, thighs, and legs without wings or back. Thighs and breasts are sold boneless and/or skinless. Dark meat (legs, drumsticks and thighs) pieces are typically cheaper than white meat pieces (breast, wings). Chicken livers and/or gizzards are commonly available packaged separately. Other parts of the chicken, such as the neck, feet, combs, etc. are not widely available except in countries where they are in demand, or in cities that cater to ethnic groups who favour these parts.

There are many fast food restaurant chains on both a national and global scale that sell exclusively or primarily in poultry products including KFC (global), Red Rooster (Australia), Hector Chicken (Belgium) and CFC (Indonesia). Most of the products on the menu in such eateries are fried or breaded and are served with french fries.

## Cooking

Raw chicken can be frozen for up to two years without significant changes in flavor or texture. Chicken is typically eaten cooked as when raw it often contains *Salmonella*. Chicken can be cooked in many ways. It can be made into sausages, skewered, put in salads, grilled, breaded and deep-fried, or used in various curries.

There is significant variation in cooking methods amongst cultures. Historically common methods include roasting, baking, broasting, and frying. Today, chickens are frequently cooked by deep frying and prepared as fast foods such as fried chicken, chicken nuggets, chicken lollipops or buffalo wings. They are also often grilled for salads or tacos. Chickens often come with labels such as "roaster", which suggest a method of cooking based on the type of chicken. While these labels are only suggestions, ones labelled for stew often do not do well when cooked with other methods.

Some chicken breast cuts and processed chicken breast products include the moniker "with Rib Meat." This is a misnomer, as it is the small piece of white meat that overlays the scapula, and is removed with the breast meat. The breast is cut from the chicken and sold as a solid cut, while the leftover breast and true rib meat is stripped from the bone through mechanical separation for use in chicken franks, for example. Breast meat is often sliced thinly and marketed as chicken slices, an easy filling for sandwiches.

Often, the tenderloin (pectoralis minor) is marketed separately from the breast (pectoralis major). In the US, "tenders" can be either tenderloins or strips cut from the breast. In the UK the strips of pectoralis minor are called "Chicken mini-fillets".

Chicken bones are hazardous to health as they tend to break into sharp splinters when eaten, but they can be simmered with vegetables and herbs for hours or even days to make chicken stock.

In Asian countries it is possible to buy bones alone as they are very popular for making chicken soups, which are said to be healthy. In Australia the rib cages and backs of chickens after the other cuts have been removed are frequently sold cheaply in supermarket delicatessen sections as either "chicken frames" or "chicken carcasses" and are purchased for soup or stock purposes.

### Freezing

Raw chicken maintains its quality longer in the freezer as compared to when having been cooked because moisture is lost during cooking. There is little change in nutrient value of chicken during freezer storage. For optimal quality, however, a maximal storage time in the freezer of 12 months is recommended for uncooked whole chicken, 9 months for uncooked chicken parts, 3 to 4 months for uncooked chicken giblets, and 4 months for cooked chicken. Freezing doesn't usually cause colour changes in poultry, but the bones and the meat near them can become dark. This bone darkening results when pigment seeps through the porous bones of young poultry into the surrounding tissues when the poultry meat is frozen and thawed. It is safe to freeze chicken directly in its original packaging, however this type of wrap is permeable to air and quality may diminish over time.

Therefore, for prolonged storage, it is recommended to overwrap these packages. It is recommended to freeze unopened vacuum packages as is. If a package has accidentally been torn or has opened while food is in the freezer, the food is still safe to use, but it is still recommended to overwrap or rewrap it. Chicken should be away from other foods, so if they begin to thaw, their juices won't drip onto other foods. If previously frozen chicken is purchased at a retail store, it can be refrozen if it has been handled properly. Chicken can be cooked or reheated from the frozen state, but it will take approximately one and a half times as long to cook, and any wrapping or absorbent paper should be discarded.

# 5

# Safely Preserving Eggs

Preserving eggs in the refrigerator for 9 months is unfathomable to most people, let alone safely *preserving eggs* that long without refrigeration. But today I'll make crystal clear how you can preserve eggs worry-free without compromising your health and taste standards and perhaps I'll even shed some light on how you came to believe that eggs can only be preserved in a refrigerator. I assure you that Kellene Bishop has not lost her mind; I'm not some maniacal bad guy accompanied by the theme music from Jaws as I write this. Just read further and you'll understand that this Preparedness Pro is no fraud but some other trusted entities may have been feeding you a fish story for the past several decades. Bottom line, if you look at how nature preserves eggs you can learn an awful lot about preserving your own regardless of whether your purchased them at your local superstore or you collected them yourself from your personal brood of hens.

When a chicken lays an egg, it has a natural coating on it known as the bloom. This bloom is a layer of protection for the egg which keeps out oxygen as well as harmful bacteria and germs. It's very, very common for those who raise chickens for their eggs to collect them and then keep them on the counter until they are ready to use—not put them in the refrigerator. And in fact, the majority of the grocery stores in the world don't put their eggs in a refrigerated environment. When they're ready to use the eggs, they either wash them, which eliminates the bloom and any bacteria or germs, or they crack them open—just as is—but take care not to do so over food that they are going to eat. Instead, they crack them into a separate dish or into a clean hand, but never over the food that they are mixing them. Also, in order to not spread the outside germs into the eggs, they never separate them with the egg shells; instead, they use their hands to separate the yolks from the whites. While it's absolutely normal for you and I to get our eggs from the refrigerated department of the grocery stores, it's actually very unusual in the rest of the world; eggs, butter, and most cheese products are usually just sitting on the shelf, not in a refrigerated display. Mind you, that doesn't mean that the U.S. is wrong in such practice, but it's important to remember that the majority of steps taken by the food industry are to avoid a lawsuit or to comply with whatever figments of grandeur the USDA wields as it creates various requirements. Such policies are not necessary to practice in your own home. Keeping that in mind, there are several steps you can take so that you can take advantage of great sales on eggs and not have to worry about whether or not they will all fit in the refrigerator.

## Preserving Eggs is Easy!

Yes, you can preserve your eggs long-term without worrying about your family getting sick. There are several methods you can use and I've written of a couple of ways previously, but my favourite one (because it's the easiest) is using mineral oil. All you have to do is warm a quarter cup of mineral oil (just about 10 seconds in the microwave will do). Set your eggs outside of the carton (because it will be hard to get them out one you start using the mineral oil). Put on some food handling gloves (I buy this at one of the warehouses). They are easy to use for safe food handling but they are also a lot less expensive than medical gloves and yet in some circumstances they can be used instead of medical gloves. They are NOT puncture resistant though, so exercise caution if using them for that purpose). Dab a little bit of the warmed mineral oil on your hands and then pick up an egg. Run your oiled hands all over the eggs, making sure to cover it completely with the mineral oil.

Don't worry if you're putting it on too thick or thin, just so long as every part of it is covered. When you're finished with an egg, put it in the egg carton, small pointed side down. A quarter cup of mineral oil should easily do 4 to 6 dozen eggs. Now, store your egg cartons in a cool, dry place. You want the temperature to be about 68 degrees for long-term storage—otherwise storing them like this in your regular room temperature is just fine for a few weeks. Remember, the eggs come out of a *warm* hen. I'm always asked if this will help the eggs keep longer in the refrigerator too. The answer is yes. You'll want to set a reminder on your calendar or cell phone to flip your eggs once a month at which time you'll simply flip the carton upside down gently so as not to break any of the eggs. Do this every month to maintain the integrity of the egg yolk.

Don't worry about an egg going bad and you not knowing it. Believe me; in this case theNOSE KNOWS. Rotten eggs smell awful. They won't be able to sneak up on you. The gas hydrogen sulphide develops as bacteria breaks down the proteins in the egg white, creating that nasty gas smell. There's another way to tell if your egg has gone bad. As oxygen gets into the egg, it creates an air bubble inside in place of moisture and carbon dioxide. This will cause the egg to float when placed in cold water (at least double the depth as the length of the egg).

Anything other than a floating egg is fine to ingest. Don't be put off by the other things that you might see in your egg. Some things may look odd to you, but don't worry, there is nothing wrong with your egg when you see the little red spots in the yolk. Those spots (usually referred to as meat spots) just mean that a blood vessel ruptured during the formation of the egg. They are diluted as the egg gets older, which means that if you actually see them, your eggs are still quite fresh. Another misinterpreted egg issue is when you see a white strand (chalazae) in the egg white, usually right next to the yolk. It reminds me of a white umbilical cord. They are only there to keep the egg yolk centred and are sometimes more obvious in some eggs than others. Even if you see an egg white that is cloudy or has a yellow or green hue to it, it doesn't mean that it's bad. In fact, quite the opposite because it indicates that the carbon dioxide just hasn't had enough time to escape the egg. Even if you end up with a fertilized egg, it's still safe to eat.

Of course all of these instructions for egg preservation only apply to eggs which you intend to cook prior to consuming. The nice thing about preserving eggs this way is that your deviled eggs will turn out better. The shells peel off of old hard-boiled eggs easier than fresh ones. In fact, the only downside I have found to preserving eggs this way is that they are pretty much useless when you want to whip up some egg whites. Old egg whites are very difficult to whip into a stiff stage once they're older than a couple of weeks. Yes, you can do this method on grocery store eggs. The grocery store egg distributors make sure to wash off the bloom and then refrigerate them. Your mineral oil essentially reapplies the bloom. Of course I always look at the dates on the egg cartons and get them as "new" as possible. This preserving eggs method will allow your eggs to stay good for 9 months and sometimes as much as 12. Obviously, rotating through your eggs is standard, so I like to number or alphabetize my eggs so that I can keep the egg cartons together in the batches that I purchased them. That way if there ever is a problem with a group of eggs, I know where they came from and can look at all of the eggs that I bought in that particular batch. You could also simply use a colour marker coding system and get the same result so long as you stick with a particular colour cycle.

### *Mineral Oil is the key to Preserving Eggs*

One last thing to mention here; the mineral oil. Mineral oil is found in the pharmaceutical sections along with Pepto Bismal, laxatives, etc. as it's commonly used to help with bowels issues. However, I only use it for this purpose and with gloves on because it is a petroleum product. Petroleum products do cause estrogen dominance in a woman's body which then causes a host of other problems in the body.

So please don't use it topically. This also goes for baby oil which is nearly 100% mineral oil except for the addition of a fragrance; though this means that you could also use baby oil in lieu of mineral oil. Now that you know that preserving eggs is easy, you can use your egg substitutes to bake with and you'll still have plenty of great eggs for those times when you just gotta have an egg over easy, deviled eggs, poached eggs on toast, homemade mayonnaise or a delectable hollandaise sauce!

A good hollandaise sauce makes the little bit of time I have to spend preserving eggs all worth it!

## Cold Cut

Cold cuts are precooked or cured meat, often sausages or meat loaves, that are sliced and usually served cold on sandwiches or on party trays. They can be bought pre-sliced in vacuum packs at a supermarket or grocery store, or they can be purchased at a delicatessen or deli counter, where they might be sliced to order.

Most pre-sliced cold cuts are higher in fat, nitrates, and sodium than those that are sliced to order, as a larger exposed surface requires stronger preservatives. In any case, the CDC (Centres for Disease Control and Prevention) advises that those over 50 reheat cold cuts to "steaming hot" 165 °F (73.9 °C) and use them within four days.

Cold cuts also may be known as lunch meats, luncheon meats, sandwich meats, cooked meats, sliced meats, cold meats and deli meats. In Commonwealth countries and the U.K., luncheon meat refers specifically to products that can include mechanically reclaimed meat, and offal. In British English, the terms *cold meats, cooked meats*, or *sliced meats* are used instead.

### *Popular*

The Spanish word for cold cut, *fiambre*, is also used in street slang to refer to a dead body (more common in Chile and Argentina), because of the word used to express low temperatures of the bodies.

In Guatemala, a cold cut is a traditional dish eaten in November. It is eaten the first and second day of the month to celebrate "El día de Todos los Santos" (All Saints' Day) and "El día de Todos los Difuntos" (All Souls' Day). There are two types: red and white.

## Meatloaf

Meatloaf is a dish of ground meat formed into a loaf shape, mixed with egg and breadcrumbs, then baked or smoked. The loaf shape is formed by either cooking it in a loaf pan, or forming it by hand on a flat baking pan. Meatloaf is usually made from ground beef, although lamb, pork, veal, venison, and poultry or a combination are also used.

The meatloaf has European origins; meatloaf of minced meat was mentioned in the famous Roman cookery collection *Apicius* as early as the 5th century. Meatloaf is a traditional German and Belgian dish, and it is a cousin to the Dutch meatball. American meatloaf has its origins in scrapple, a mixture of ground pork and cornmeal served by German-Americans in Pennsylvania since Colonial times. However, meatloaf in the contemporary American sense did not appear in cookbooks until the late 19th century.

### *Dish*

Many meatloaf recipes are interchangeable with meatball recipes, the distinction coming from shape and from the accompaniments or choice of sauce. Sometimes a form of tomato sauce or ketchup is also mixed with the meat before baking.

The ground meat may be mixed with a binder such as eggs, breadcrumbs and TVP, small pieces of bread soaked in milk (or red wine or another liquid), or wheat germ, finely ground almonds, or oatmeal, to make it more dough-like.

Salt, spices and herbs (such as parsley), as well as chopped vegetables (such as onions, green peppers and celery), may also be added. Some recipes add strips of bacon on top. Meatloaf may be filled with eggs, cheese, vegetables, herbs, mushrooms, or a combination of these ingredients. A regular oven is not required; meatloaf can be cooked in a slow cooker or in a microwave oven.

## International Variations

***Austria:*** The Austrian meatloaf version is called *Faschierter Braten*. Most of the time it is not filled (e.g., in Germany), but it is wrapped in ham before baking it. Often it is served with mashed potatoes (when warm) or with sauce cumberland (when cold).

***Belgium:*** The Belgian version of meatloaf is called *pain de viande.* It is usually served warm and can then be served with various sauces, but can also be eaten cold with a loaf of bread.

***Bulgaria:*** *Rulo Stefani.* The Bulgarian *rulo Stefani* meatloaf is similar to the Hungarian *Stefánia* meatloaf, with hard-boiled eggs in the middle.

***Czech Republic:*** In the Czech Republic, meatloaf is referred to as *sekaná* ('chopped'). It is optional to put hard boiled eggs, gherkins, or wienerwurst inside.

***Denmark:*** Danish meatloaf is called *forloren hare* 'mock hare' or *farsbrød* 'ground-meat bread' and is usually made from a mixture of ground pork and beef with strips of bacon or cubed bacon on top. It is served with boiled or mashed potatoes and brown sauce sweetened with red currant jelly.

***Finland:*** Finnish meatloaf is called *lihamureke*. It is completely based on the basic meatball recipe. The only spices used are salt and pepper. It is not customary to stuff *lihamureke* with anything. The usual side dish is mashed potatoes, and *lihamureke* is usually served with brown sauce.

***Germany:*** In Germany, meatloaf is referred to as *Hackbraten, Faschierter Braten* or *Falscher Hase* 'mock hare'. In some regions it often has boiled eggs inside.

***Greater Middle East:*** *Kafta* or *kofta* is a similar dish which the mixture can be made into hamburgers and kebabs. It usually has parsley in it.

***Greece:*** In Greece, meatloaf is referred to as *rolo* (Ñïëü) and it is usually filled with hard boiled eggs, although several other variations exist.

***Hungary:*** Stefania meatloaf (Hungarian: *Stefánia szelet*) or Stefania slices are a type of Hungarian long meatloaf baked in a loaf pan, with 3 hard boiled eggs in the middle, making decorative white and yellow rings in the middle of the slices.

***Italy:*** In Italy, meatloaf is called *polpettone* and can be filled with eggs or ham and cheese.

***Jewish cuisine:*** In Jewish cuisine, meatloaf is called *Klops* (Hebrew) and can be served cold or hot. It is sometimes filled with whole boiled eggs. The source of the word might be German, *Klops*, meaning meatball.

***Palestine/Lebanon:*** "Kafa" or "Kofta"is one of the middle eastern version of meatloaf. It is usually beef or lamb mixed with onions, parsley, and various spices flattened into a square loaf and covered in either tomato sauce or tahini sauce. It is baked with the sauces and enjoyed with rice and potatoes (usually baked with the kafta). Kafta or Kofta is also turned into kababs and grilled during the summer months.

***Macedonia:*** *Rolat* is a similar dish to the chiefly Arab, though also Persian and South-Asian, kofta. Ground beef is rolled and cooked until brown. It can be cooked with vegetables and various sauces.

***Mongolia:*** (khoochmal) is served with mashed potatoes cooked over the ground meat.

***Philippines:*** There is a meatloaf dish called *embotido* made of well seasoned ground pork, minced carrots, sausages, and whole boiled eggs. The meat is molded into a roll with the sausages and hard boiled eggs set in the middle. It is then wrapped in aluminum foil (historically, banana leaves) and steamed for an hour. The cooked *embotido* may be stored in freezers. It is usually served fried and sliced for breakfast.

***Romania:*** In Romanian cuisine, there is a meatloaf dish called *drob*, similar to other minced meat dishes in the region like the Bulgarian *Rulo Stefani* or the Hungarian *Stefánia* meatloaf, the major difference being that it is always made with lamb organs (or a mixture of lamb organs and pork or veal) and the hard boiled eggs in the centre of the *drob* are optional.

***South Africa:*** One of South Africa's most popular dishes is a form of meatloaf called bobotie. Food in South Africa varies by region, and bobotie is considered a Cape Town or Western Cape dish, but is so popular throughout the country that it is sometimes considered one of South Africa's signature dishes. It is a slightly sweet meatloaf flavored with curry spices, dried appricots and almonds, and is topped with whipped eggs. It arrived in South Africa several hundred years ago from the East Indies and became one of the signature dishes of Cape Town's "coloured" community.

***United Kingdom:*** In the UK there are regional meatloaf dishes known as Haslet.

***United States:*** In 2007, meatloaf was voted the seventh-favourite dish in the United States according to *Good Housekeeping*.

During the Great Depression, cooking meatloaf was a way to stretch the food budget for families, using an inexpensive type of meat and other ingredients as leftovers; along with spices, it was popular to add cereal grains to the meatloaf to stretch the meat. The tradition lives on with the merits of producing a lower-fat dish with superior binding and consistency.

The meatloaf is typically eaten with some kind of sauce or relish. Many of these recipes call for pasta sauce or tomato sauce to be poured over the loaf to form a crust during baking. The tomato-based sauce may be replaced with simple brown gravy or onion gravy, but the meatloaf is prepared in a similar manner. Barbecue sauce, tomato ketchup, or a mixture of both tomato ketchup and mustard may also be used. American meatloaf may be garnished with plain ketchup or a "meatloaf sauce" consisting of ketchup and Brown sugar mixed.

Another variety of meatloaf is prepared by frosting it with mashed potatoes, drizzling it with a small amount of butter, and browning in the oven.

The meatloaf is normally served warm as part of the main course, but can also be found sliced as a cold cut. Meatloaf can also be considered a typical comfort food and is served in many diners and restaurants today.

***Vietnam:*** The Vietnamese meatloaf version is called giò. It's boiled rather than baked or smoked. There are many versions of giò that differ by the materials used.

## Meatball

A meatball is made from an amount of ground meat rolled into a small ball, sometimes along with other ingredients, such as breadcrumbs, minced onion, spices, possibly eggs and herbs. Meatballs are usually prepared and rolled by hand, and are cooked by frying, baking, steaming, or braising in sauce.

There are many types of meatball recipes using different types of meats and spices, including vegetarian and fish alternatives, and various methods of preparation.

The ancient Roman cookbook *Apicius* included many meatball-type recipes. From the Balkans to India, there is a large variety of meatballs in the kofta family.

### Meatballs Across Various Cultures

- In Afghanistan, meatballs are used as a traditional dish with homemade soups, or are made with a tomato based sauce that may include some plum seeds to increase tartness and is served with bread or rice which is called Kofta-Chelou. Now meatballs are also grilled on top of pizza.
- Albanian fried meatballs (*qofte të fërguara*) include feta cheese.
- In Austria, fried meatballs are called *Fleischlaibchen* or *Fleischlaberl*.
- In Belgium, meatballs are called *ballekes* or *bouletten* in Flanders, and are usually made of a mixture of beef and pork with breadcrumbs and sliced onions. Many other variations exist, including different kinds of meat and chopped vegetables.
- In Bulgaria meatballs are called *kyufte* and are typically made from ground beef or pork, or a mix of the two. They can be shallow fried or grilled and often contain diced onions and soaked bread. They are a very popular dish.
- Chinese meatballs (specifically, a dish common in Shanghai cuisine) are most often made of pork and are usually steamed or boiled, either as-is, or with the addition of soy sauce. Large meatballs, called lion's heads, can range in size from about 5 cm to 10 cm in diametre. Smaller varieties, called pork balls, are used in soups. A Cantonese variant, the steamed meatball, is made of beef and served as a dim sum dish. A similar dish is called the beef ball, and the fish ball is yet another variety made from fish ground into a fine paste. In northern China, irregular balls made from minced meat and flour are often deep-fried and eaten for special occasions.
- Danish meatballs are known as *frikadeller* and are typically fried, and they are usually made out of ground pork, veal, onions, eggs, salt and pepper; these are formed into balls and flattened somewhat, so they are pan ready.
- In Finnish cuisine, meatballs (*lihapullat*) are made with ground beef or a mix of ground beef and pork, or even with ground reindeer meat, mixed with breadcrumbs soaked in milk and

finely chopped onions. They are seasoned with white pepper and salt. Meatballs are traditionally served with gravy, boiled potatoes (or mashed potatoes), lingonberry jam, and sometimes pickled cucumber.

- In Alsacia, France, meatballs are known as Fleischkiechele. They are made of beef, pork, onions, bacon, eggs and bread. They are served plain or with cream sauce.
- In Germany, meatballs are mostly known as *Frikadelle*, *Fleischpflanzerl*, *Bulette* or *Klopse*. A very famous variant of meatballs are *Königsberger Klopse*, which contain anchovy or salted herring, and are eaten with caper sauce.
- In Greece, fried meatballs are called *keftédes* and usually include within the mix bread, onions and mint leaf. Stewed meatballs are called *yuvarlákia* (from the Turkish word *yuvarlak*, which means "round") and usually include small quantities of rice.
- In Hungary, as well as territories from neighbouring countries where Hungarian is spoken, a meatball is called *fasirt* or *fasirozott* probably coming from Austrian German *faschierte Laibchen*. Also the *májgombóc* (liver dumpling) is popular in soups.
- In Iran, several types of meatballs are consumed. If they are cooked in a stew, they are called *kufteh*. If they are fried (typically small meatballs), they are called *kal-e gonjeshki* (literally "sparrow's head"). Both types are consumed with either bread or rice. Typically, herbs are added, and for *kufteh*, usually the meatball is filled with hard boiled eggs or dried fruits. There are several (at least 10) types; the most famous is "kufteh tabrizi", traditionally from Tabriz in northwestern Iran.
- In Indonesia, meatballs are called *bakso* which are usually served in a bowl, like soup, with noodles, beancurd (tofu), egg, *siomay*/steamed meat dumpling, and crispy wonton. They have a consistent homogeneous texture. *Bakso* can be found all across Indonesia, but the most popular are *bakso Solo* and *bakso Malang* (named for the city from which it originates). In Malang, *bakso bakar* (roasted *bakso*) is also popular. As most Indonesians are Muslim, generally it is made from beef or is mixed with chicken.

- In Italy, meatballs are generally eaten as a main course or in a soup. The main ingredients of an Italian meatball are: beef and or pork and sometimes poultry, salt, black pepper, chopped garlic, olive oil, Romano cheese, eggs, bread crumbs and parsley, mixed and rolled by hand to a golf ball size. In the Abruzzo region of Italy, especially in the Province of Teramo, the meatballs are typically the size of marbles, and are called *polpettine.*
- The Japanese hamburger steak, *hanbâgu,* is typically made of ground beef, milk-soaked panko (bread crumbs) and minced, sauteed onions. They are typically eaten with a sauce made from ketchup and Worcestershire sauce. Chinese style meatballs are also popular.
- In the Netherlands, meatballs are called *gehaktbal,* and are often served with boiled potatoes and vegetables.
- In Norway, meatballs are called *kjøttboller* (lit. "meatbuns"). The influence from Swedish meatballs is such that they are even often referred to as "köttbullar", though usually jokingly or because the meatballs were actually purchased in Sweden, which is common in areas close to the Swedish border. Meatballs come in a few different types, all typically small, and the international influence is great, perhaps the greatest from Sweden and Spain. They are usually eaten with potatoes or pasta, or both. Some common additions are various vegetables, ketchup, various spices etc. "Kjøttkaker" (lit. "meatcakes") is a much larger and different related dish, and is perhaps more traditional and also common, which is much larger in size and made of different things in a different way (and the two should not be confused). The latter is often served with potatoes and peas (either could be mashed). Kjøttboller is typically fried, a process which takes only very few minutes because of their size, whereas kjøttkaker are typically part of a mix which includes a brown sauce and often potatoes.
- In the Philippines, meatballs are called *almondigas* or *bola-bola,* and usually served in a soup with rice vermicelli called *misua,* toasted garlic, squash and pork cracklings.
- In Poland, they are called *pulpety* or *klopsy* (singular *pulpet*; *klops*), and *pulpeciki* ("little *pulpety*"), and are usually served cooked with a variety of sauces (such as tomato or a kind of

gravy thickened with flour, as well as forest mushroom sauce) with potatoes, rice or all sorts of *kasza*. *Pulpety* or *klopsy* are usually made from seasoned ground meat with onion and mixed with eggs and either breadcrumbs or wheat rolls soaked in milk or water. Fried *pulpety* are larger than typical cooked ones. They can be round or flat in shape. The latter, in many countries, would be considered a cross between a meatball and a hamburger. The fried variety is called *mielony* (short for *kotlet mielony*, literally "minced cutlet"), and its mass-produced version (as well as the one served in bars, etc.) is the subject of many jokes and urban legends about what is used to produce it.

- In Portugal and Brazil, meatballs are called *almôndegas*. These are usually served with tomato sauce and pasta.
- In Romania, meatballs are called *chiftele* and *pârjoale* and are usually deep fried and made with pork or poultry, moistened mashed potatoes and spices. *Chiftele* are flat and round and contain more meat. A variant mixing rice meatball is used for sour soup, making *ciorbã de peri°oare*.
- In Spain and Latin America, meatballs are called *albóndigas*, derived from the Arabic *al-bunduq* (meaning hazelnut, or, by extension, a small round object). *Albóndigas* are thought to have originated as a Berber or Arab dish imported to Spain during the period of Muslim rule. Spanish *albóndigas* can be served as an appetizer or main course, often in a tomato sauce, while Mexican *albóndigas* are commonly served in a soup with a light broth and vegetables.
- In Sweden, *köttbullar* (meatballs) are made with ground beef or a mix of ground beef, pork and sometimes veal, sometimes including breadcrumbs soaked in milk, finely chopped (fried) onions, some broth and often including cream. They are seasoned with white pepper or allspice and salt. Swedish meatballs are traditionally served with gravy, boiled potatoes, lingonberry jam, and sometimes fresh pickled cucumber. Traditionally, they are small, measuring one inch in diametre. In the United States, there are a number of variations, based on the assimilation of Swedes in the Midwest.
- In Turkey, meatballs are called Köfte and are extremely popular, there are at least 50 different versions. Meatballs in Turkey are usually made with ground lamb or a mix of ground beef

and lamb. Most popular ones are Ýnegöl Köfte, Ýzmir Köfte, Þiþ Köfte, Kadýnbudu Köfte and Akçaabat Köftesi.

- In the United Kingdom, faggots are a type of spicy pork meatball. A faggot is traditionally made from pig heart, liver and fatty belly meat or bacon minced together, with herbs added for flavouring, and sometimes breadcrumbs.

***Figure:*** *A bowl of* Bún mÌc *in Vietnam,* bún *is rice vermicelli* mÌc *means pork-ball*

- In the United States, meatballs are commonly served with spaghetti as in spaghetti and meatballs, a dish in Italian American cuisine, assimilated from Italian immigrants coming from southern Italy in the early 19th century. Over time, the dishes in both cultures have drifted apart in similarity. In the southern United States, venison or beef is also often mixed with spices and baked into large meatballs that can be served as an entree. Another variation, called "porcupine meatballs" are basic meatballs often with rice in them.
- In Vietnam, meatballs can be used as an ingredient in *phß, hç ti¿u*. It is also common to cook meatballs in tomato sauce, and finely chopped spring onion and peppers are added before serving. In *bún ch£* (a specialty Vietnamese rice noodle), meatballs are grilled to be *ch£* and served with *bún* (rice

noodles) and dipping sauce (based on fish sauce seasoned with rice vinegar, sugar, garlic, and chili). *Xíu M¡i* is a pork meatball in a tomato sauce often served with a baguette.

## Kofta

Kofta is a Middle Eastern and South Asian meatball or dumpling. The word *kofta* is derived from Persian *kûfta*: In Persian, ˜æÝÊä (kuftan) means "to beat" or "to grind" or *meatball*. In the simplest form, koftas consist of balls or fingers of minced or ground meat – usually beef or lamb – mixed with spices and/or onions. The vegetarian variety like *lauki kofta, shahi aloo kofta, malaai kofta* are popular in India.

The meat is often mixed with other ingredients such as rice, bulgur, vegetables, or eggs to form a smooth paste. Koftas are sometimes made with fish or vegetables rather than meat, especially in India. They can be grilled, fried, steamed, poached, baked or marinated, and may be served with a rich spicy sauce. Variations occur in North Africa, the Mediterranean, Central Europe, Asia and India. According to a 2005 study done by a private food company, there were 291 different kinds of kofta in Turkey, where it is very popular. In Arab countries, kufta is usually shaped into cigar-shaped cylinders.

Early recipes (included in some of the earliest known Arabic cookbooks) generally concern seasoned lamb rolled into orange-sized meatballs, and glazed with egg yolk and sometimes saffron. This method was taken to the west and is referred to as gilding, or endoring. Many regional variations exist, notable among them the unusually large Iranian Kufteh Tabrizi, having an average diametre of 20 cm (8 in).

*Koftas* in South Asian cuisine are normally cooked in a spicy curry and sometimes with whole pre-boiled eggs. Sometimes the eggs are encased in a layer of the spicy kofta meat so that the final product resembles an Indian Scotch egg. These kofta dishes are very popular with South Asian families and are widely available from many Indian restaurants. In West Bengal, India and Bangladesh, koftas are made with prawns, fish, green bananas, cabbage, as well as minced goat meat.

## General Hygiene Rules for Facilities, Equipment and Personnel in the Meat Industries

It is essential that all meat-processing operations, whether slaughtering, cutting or further processing, be carried out in a clean area and, as much as possible, that the products be protected from contamination from all sources.

When meat-processing operations are carried out within a facility specifically built and maintained for meat processing, sources of contamination can be much more easily and adequately controlled. The following requirements are considered essential to good sanitary preparation of meat and meat products.

### Facilities

*Floors.* Brick, tile, smooth concrete or other impervious, waterproof materials are suitable for floors. In some areas wooden floors will suffice if they are tight, smooth, in good repair and properly maintained. Wooden floors are not suitable in areas where slaughtering or curing takes place and meat juices and moisture collect.

*Drains.* To carry away waste liquids, there should be sufficient drains of the proper size that are correctly located, trapped and vented. All floors should be sloped towards the drains. Generally for adequate waste disposal, one drain is needed for each 18 $m_2$ of floor space in slaughtering areas, and one drain for each 46$m_2$ in processing and other areas.

*Walls.* Glazed tile, smooth cement plaster, rustproof metal panels and smooth plastic panels that are properly caulked are all acceptable for walls in processing and refrigerated areas because they can all be effectively cleaned and sanitized. Other materials are also acceptable if they can be satisfactorily cleaned. In no instance should walls be made of materials that absorb moisture or other liquids. Ceilings must be tight, smooth and free from any scaling that may fall into the meat products, and should also be of moisture-resistant materials. All light bulbs should be covered with unbreakable material to prevent broken pieces from falling into the product.

*Doors and doorways.* All doorways through which the product must pass, whether suspended on rails or lying on hand trucks, should be wide enough to ensure that the meats never touch the doorways risking contamination. Wooden doors and doorways should be covered with metal with tightly soldered seams.

*Water supply.* Whether from individually owned and controlled sources such as wells or streams or from a municipal system, the water supply must be potable and abundant cold and hot water must be distributed to all parts of the operation.

*Lighting.* In all areas where products are critically examined during sanitary control or for cleanliness, 50-foot candles of light should be provided. For adequate visibility 20-foot candles of light should be provided wherever any processing occurs. In all other areas, such as dry storage, there should be sufficient light to keep the area orderly and sanitary.

*Refrigeration.* The main purpose of refrigeration is to cool the meat down after slaughter and to maintain it in a chilled state for shorter or longer storage periods and for cutting and further processing. If frozen storage is provided and utilized, it should be maintained at the lowest possible temperature for maximum shelf-life. Minus 18° to -12°C is satisfactory freezer storage; however, large quantities of product must either be quickfrozen prior to storage or thinly spread out to facilitate freezing. It is also recommended that all rooms where meat is processed, except in the slaughter and cooler storage areas, should be maintained at a temperature of about 12°C. In facilities where no refrigeration or cooling is furnished in processing areas, the handling of meat products is possible if all equipment contacting the products is throughly cleaned and sanitized from time to time (recommended every four hours). Frequent cleaning is necessary because in warmer temperatures bacteria multiply rapidly and the risk of product contamination increases.

### Equipment

The equipment needed for converting livestock into meat products need not be elaborate and expensive. The amount of equipment will depend on the slaughtering and processing procedures employed. If possible, all equipment should be made of stainless steel or plastic, be rust resistant and easily cleaned and sanitized.

All equipment should be constructed of stainless steel, galvanized steel, aluminium or approved plastic. Wooden tables are not acceptable because wood absorbs meat juices and fats and cannot be thoroughly cleaned. Hardwood cutting-boards maintained smooth and free from checks and cracks may be used. Cutting tables covered with other than hard plastic are not acceptable for contact with meat. All other equipment should be of the type that can be taken apart and thoroughly cleaned. Any stationary equipment must be located far enough from walls to permit proper cleaning around and under it.

In all areas there should be conveniently located foot-pedal or kneeoperated wash-basins with hot and cold water, soap and disposable towels. In slaughtering areas, lavatories should be convenient to the dressing operations. Hot-water containers, either electric or steam-heated to 82°C, should be available for sanitizing tools contaminated with diseased material or other filth during dressing.

Rails must be located high enough to prevent meat from touching the floor. For beef carcasses, the minimum height for rails should be 3.4 metres, while 2.4 metres is sufficiently high for small livestock such as goats, hogs and sheep. Rails should also be far enough away from fixed objects and walls to avoid contact.

### Personnel Hygiene

Probably as important as anything in the production of clean, wholesome, unspoiled products is the attitude of the workers towards cleanliness. Personnel with clean hands, clothing and good hygienic practices are absolutely essential to the production of high-quality foods.

All clothing should be clean, in good repair and made of washable material. Street clothing should be covered with coats or gowns while handling exposed product. White or light-coloured clothing is most desirable and garments that become soiled or contaminated should be changed when necessary.

All persons working with exposed meat products should have their hair under control, either completely covered with a clean cap or hat or confined by a hairnet to prevent hair from falling into products.

Safety devices such as aprons, wrist guards and mesh gloves must be made of impervious material, clean and in good repair. At no time should leather aprons, wrist guards or other devices be worn unless clean, washable coverings are used over them. Light-coloured rubber or plastic gloves may be worn by product handlers only if clean and in good repair.

No person working with meats should wear any kind of jewellery, badges or buttons that may come loose and be accidently included in the product.

Shoes and boots should be worn at all times and should be appropriate for the operations being conducted. They should also be made of impervious materials. Any aprons, knives and footwear that become contaminated during operations should be routinely cleaned in areas or facilities provided for that purpose.

No cloth twine, belts or other similar materials should be used to cover implement handles or used in other places where they may harbour filth and serve as a ready source of product contamination.

All unsanitary practices should be avoided by meat handlers. No one should smoke or use tobacco in areas where edible products and ingredients are handled, prepared or stored, or where equipment and utensils are cleaned.

When handling edible products, scratching the head, placing fingers in or around the nose or mouth, sneezing or coughing on the product should never occur. Workers must also guard against contaminating products from localized infections or sores.

Workers can contaminate carcasses and meat through handling, coughing and sneezing. This may cause rapid spoilage of the meat or, more seriously, food poisoning. Coughs and sneezes are a particularly effective way of transmitting bacteria to meat. Transfer of faecal matter either of animal or human origin to the meat is particularly hazardous. Most contamination on the hands of workers in slaughter floors with faecal matter comes from the hides and fleeces.

Hands should be washed frequently to remove all visible soiling. Stainless-steel sinks without plugs should be conveniently accessible to all workers. Water should be supplied at approximately 43°C to a simple tap which is foot- or knee-operated. Liquid disinfectant soap and paper towels should be available. Particular attention should be paid to cleaning under the fingernails. Hands should also be thoroughly washed after using the toilet, smoking, coughing or sneezing, handling money, garbage or soiled or infected material.

All precautions should be taken to prevent product contamination by visitors or other persons who are simply passing through the work area.

### *Routine Cleaning of Rooms and Equipment*

The floors should be kept clear of all debris, such as hooves and horns, in slaughterhalls or other inedible parts or fat and meat particles in cutting, processing and by-product handling areas, and must be frequently washed down.

At the end of each day a thorough cleaning programme should be followed. All matter should be removed from floors, platforms, gullies, etc., followed by a thorough hosing down of walls, floors and all surfaces to loosen dirt.

Finally a strong cleaning solution should be applied and left for a while before being rinsed off. A thorough inspection should be made afterwards and any areas remaining soiled should be cleaned again.

In order to maintain the cleanest possible products a standard cleaning routine of the equipment should be established. Initially all large pieces of refuse material should be scraped or swept together and disposed of. Follow-up should include scrubbing of the equipment using brushes and a soap or detergent and a complete sanitizing with hot water at 82°C and an approved chlorine or iodine rinse. Finally, a coating of light mineral oil can be applied to metal equipment, particularly that not fabricated of stainless steel, to prevent rust.

# 6

# General Hygiene Principles for Meat Handling

Current recommendations for handling all meat products are to keep them clean, cold and covered in order to maintain quality and protect against food poisoning and disease. Generally contamination occurs when the product comes into contact with dirty hands, clothing, equipment or facilities.

If the product is kept clean there will be little or no contamination by microorganisms whether bacteria, yeasts, moulds, viruses or protozoa or by helminths and poisonous chemicals.

## Effect of Bacterial Growth on The Shelf-life of Meat and Meat Products

The total viable count of bacteria (TVC) expressed as organisms/$cm^2$ or as organisms/g on fresh meat or a meat product sets a limit to its shelf-life. Meat will "spoil" with TVC at $10^6/cm^2$ because of off-odours. Slime and discolouration appear at $10^8/cm^2$. The main factors determining the time taken for the TVC to reach these levels are the initial count due to contamination during slaughtering and processing, further contamination during storage, temperature, pH and relative humidity. An example of how the level of contamination affects shelf-life is shown.

After cleanliness, keeping meat products cold is the second most important requirement in order to achieve a desirable shelf-life. Microorganisms rapidly proliferate at elevated temperatures and slime development is a definite visual sign of microbial growth. The importance of temperature in the control of microbial growth is shown.

***Table:** Effect of initial contamination on the storage life of lean beef*

| Initial bacterial count (org./cm²) | Days at 0°C before slime development |
|---|---|
| 100 000 | 8 |
| 10 000 | 10 |
| 1 000 | 13 |
| 100 | 15 |
| 10 | 18 |

***Table:** Relationship between storage temperature and slime development*

| Storage temperature (°C) | Days before slime develops |
|---|---|
| 0 | 10 |
| 1 | 7 |
| 3 | 4 |
| 5 | 3 |
| 10 | 2 |
| 16 | 1 |

Under ideal conditions bacteria double in number every 20 minutes. A single bacterium multiplies to over one million in less than seven hours:

| Time | Number of bacteria |
|---|---|
| 12.00 | 1 |
| 12.20 | 2 |
| 12.40 | 4 |
| 13.00 | 8 |
| 14.00 | 64 |
| 15.00 | 512 |
| 16.00 | 4 096 |
| 17.00 | 32 768 |
| 18.00 | 262 144 |
| 18.40 | 1 048 576 |

Bacteria relevant to meat, meat products and other food are divided into three groups according to the temperature range within which they can grow: mesophiles 10–45°C, psychrophiles 0–28°C and psychrotrophs 10– 45°C, or slow growth at 0–10°C. Mesophiles will not grow below 10°C but psychrotrophs, of which Pseudomonas are the more important, will grow down to 0°C. The nearer to 0°C the storage temperature the slower the growth of the spoilage bacteria and the longer the shelf-life.

Some bacteria cause product spoilage, others cause food poisoning. The former limit product shelf-life but the latter cause illness. Almost all foodpoisoning bacteria are mesophiles so refrigeration below 10°C offers good protection. Many mesophiles cause spoilage, but since meat is refrigerated most spoilage is due to psychrophiles. Storing meat at temperatures close to 0°C will inhibit the growth of pyschrotrophs. Shelf-life will be extended by avoiding contamination through good hygiene practices.

Aerobic bacteria have an absolute requirement for oxygen which limits their growth to the meat surface. Anaerobic bacteria grow within the meat as they need the absence of oxygen. Facultative anaerobes can grow slowly within oxygen but grow better in its presence. Food-poisoning bacteria are anaerobes and facultative anaerobes. The most important spoilage bacteria (*Pseudomonas* spp.) are aerobic.

Water is required by micro-organisms so reducing the water available below the optimum level will prolong shelf-life. If meat is stored at a relative humidity (RH) below 95 percent, moisture will be lost from the surface. Since most spoilage bacteria, being aerobic, can grow only on the surface, this surface drying will extend the shelf-life. Moulds (fungi) are able to grow in drier conditions than bacteria so that desiccation has a selective effect on microbial growth.

Meat pH is the level of acidity in meat. Stored sugars are broken down to lactic acid. In living muscle it is near 7.0 (above this is alkaline, below is acid). It falls to 5.4–5.6 within 24 hours. High final-pH values result when animals are exhausted at slaughter, for instance because of fighting in lairage or transport. Spoilage bacteria multiply rapidly on high-pH meat and shorten the shelf-life. Exhausted animals should be rested before slaughter. A high TVC resulting from severe contamination during slaughter or processing will shorten the shelf-life even in ideal conditions. It also indicates poor hygiene so that contamination with food-poisoning bacteria is likely.

### *Effect of Contamination on Sensoric Properties of Meat and Meat Products*

Aerobic spoilage by bacteria and yeasts usually results in slime formation, undesirable odours and flavours (taints). Colour changes, rancidity, tallowy or chalky flavours from the breakdown of lipids may also occur. Colour changes as a result of pigment oxidation may be grey, brown or green discolouration. Aerobic spoilage by moulds results in a sticky surface, musty odours, alcohol flavours and creamy, black or green discolouration.

Anaerobic spoilage which occurs either within the meat or on the surface in sealed containers where oxygen is absent or very limited is marked by a souring due to the production of organic acids and gases.

### Food Poisoning

Food poisoning may be due to infection or intoxication. Infection is caused by the consumption of live bacteria which multiply in the body producing characteristic symptoms. Intoxication is due to toxins in food produced by bacteria before the food was eaten. Toxins are chemical compounds which may linger in food with no microbes growing in it, and are therefore very dangerous.

Salmonellae are facultative anaerobes which cause infectious food poisoning. Ten or 20 cells of *Salmonella typhi* are sufficient to cause typhoid but 10 000 to 100 000 cells of other species may be necessary to cause an infection. Some are host-specific affecting the animal from which the meat was produced but failing to cause infection when consumed by man. Typical symptoms of salmonellosis include diarrhoea, fever and vomiting. The illness may last one to 14 days after a 12 to 24-hour incubation period. Victims may excrete the bacteria for weeks after the symptoms subside. Poor personal hygiene will cause contamination of meat.

*Staphylococcus aureus* is a facultative aerobe that causes intoxication. It lives in the nose, throat, hair and skin and on animal hides. Meat is contaminated by handling and by sneezing or coughing. Minute amounts of the toxin will cause illness, which starts within one to eight hours of eating poisoned food. Nausea, vomiting and shock may last for one to two days.

On rare occasions it is fatal. This bacterium does not produce off-odours or spoilage so it cannot be easily checked. Refrigeration will control its growth. Cooking may destroy the bacteria but not the toxin as it is heat stable. It is particularly troublesome in cooked cured meats, normally as a result of recontamination after the curing process in subsequent handling, for instance during slicing.

*Clostridium botulinum*, an anaerobe, produces the toxin botulin, one of the most poisonous substances known. This attacks the central nervous system causing death by respiratory paralysis. Dormant cells occur everywhere in the soil, fish, animals and plants.

High-moisture, low-acid, low-salt conditions at above 3°C favour growth and toxin production. Control measures must destroy spores or prevent growth and toxin formation. Botulism is usually due to undercooking processed meats. Pressure-cooking will give commercial sterility. Pasteurization (heating to 70°C) and adding salt (NaCl) and sodium nitrite ($NaNO_2$) is used for canned ham. Refrigeration (0–10°C) is essential for vaccum-packed meats. Frozen storage prevents growth.

*Clostridium perfringens*, an anaerobic bacterium, is a common cause of food poisoning but is rarely fatal. It grows well in warm meats so is usually found in left-over meats that have not been kept chilled and not been reheated to 70°C to kill the bacteria present. The main symptoms are diarrhoea and weakness which last for 12 to 24 hours after an incubation period of eight to 20 hours.

## Techniques and Hygiene Practices in Slaughtering and Meat Handling

### *Equipment*

Slaughtering equipment, particularly for smaller-scale operations, need not be elaborate and expensive. The amount of equipment will depend on the slaughtering procedures employed. If possible, all equipment should be made of stainless steel or plastic, be rust resistant and easily cleaned and sanitized. Equipment which does not get in contact with the meat (e.g. overhead rails, working platforms, knocking pen) is usually made of galvanized steel.

Basic equipment needed for the slaughtering operation:

- stunning gun, electrical head tongs or simple stunning equipment for direct blow
- knives:

sticking - 15 cm sharpened on both sides

skinning - 15 cm curved

- a sharpening steel
- oil or water sharpening stone
- scabbard and belt for holding knives

- meat saw - hand or electric and cleaver
- block and tackle or chain hoist strong enough to hold the weight of the animal to be slaughtered
- pritch, chocks or skinning rack (dressing cradle)
- a strong beam, tripod or track 2.4 to 3.4 m from floor
- spreader - gambrel or metal pipe
- several buckets
- working platforms
- scalding barrel or tank
- pot, barrel or system for boiling water
- bell scrapers
- solid scraping table or platform
- thermometre registering up to 70°C
- hog or hay hook
- torch or flame for singeing

The last seven items indicate additional equipment required when hogs are scalded and scraped rather than skinned.

Useful additional equipment:

- knocking pen
- bleeding hooks (for vertical bleeding)
- blood-catching trough
- wash trough (tripe) Sanitation of hands and tools:
- hand wash-basin
- implement sterilizers

Means should be available to clean thoroughly all equipment coming into contact with carcasses or meat. Implement sterilizers are stainless-steel boxes holding hot (82°C) water, shaped to suit particular equipmentknives, cleavers, saws, etc. Knife sterilizers should be placed in positions where every operator who uses a knife has immediate access. Handles as well as blades must be sterilized. Each operator should have at least two knives etc., one to use while the other sterilizes.

Failure to sterilize all knives and equipment regularly will result in carcass contamination. Bacteria will be transferred from the hide to the carcass and from carcass to carcass.

## Treatment of Livestock Before Slaughter and Its Impact on Meat Quality

Stress in its many forms, e.g. deprivation of water or food, rough handling, exhaustion due to transporting over long distances, mixing of animals reared separately resulting in fighting, is unacceptable from an animal welfare viewpoint and should also be avoided because of its deleterious effects on meat quality. The most serious consequence of stress is death which is not uncommon among pigs transported in poorly ventilated, overcrowded trucks in hot weather.

From loading on the farm to the stunning pen animals must be treated kindly, and the lorries, lairages and equipment for livestock handling must be designed to facilitate humane treatment. Stress immediately prior to slaughter, such as fighting or rough handling in the lairage, causes stored glycogen (sugar) to be released into the bloodstream.

After slaughter this is broken down in the muscles producing lactic acid. This high level of acidity causes a partial breakdown of the muscle structure causing the meat to be pale, soft and exudative (PSE). This condition is mostly found in pigs.

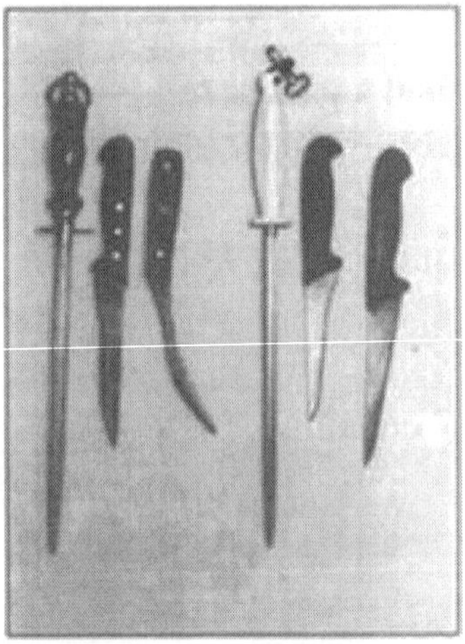

***Figure:*** *Plastic (right) handles are more hygienic than wooden (left) for knives and other equipment. Note the excessively worn knife (third from left) which should be discarded*

Long-term stress before slaughter such as a prolonged period of fighting during transport and/or lairage leads to exhaustion. The sugars are used up so that less is available to be broken down and less lactic acid is produced.

The reduced acidity leads to an abnormal muscle condition known as dark, firm and dry (DFD) in pigs or dark cutting in beef. The condition is rarer in lamb. Such meat has a high pH (above 6.0) and spoils very quickly as the low acidity favours rapid bacterial growth.

### Handling Animals During Transport and Lairage

An electric goad should be used rather than a stick or tail-twisting not only to avoid stress but also to prevent carcass bruising. Grabbing sheep by the fleece also causes bruising.

To avoid fighting, animals not reared together must not be mixed during transport and lairage. Load and unload using shallow stepped ramps to avoid stumbles. Trucks should be neither over- nor underloaded. Overloading causes stress and bruising due to crushing. Underloading results in animals being thrown around and falling more than necessary. Drivers should not corner at excessive speed and must accelerate and decelerate gently.

The lairage should have small pens. Corridors must curve and not bend sharply so that stock can see a way forward. Stock must not be slaughtered in sight of other stock. Plenty of clean water must be available. The lairage must be well lit and ventilated. Do not hold stock in lairage for more than a day. Only fit, healthy stock may be slaughtered for human consumption.

Fasting before slaughter reduces the volume of gut contents and hence bacteria and therefore reduces the risk of contamination of the carcass during dressing. It is usually sufficient for the animals to receive their last feed on the day before slaughter. Stock should have a rest period after arrival at the slaughterhouse. However, long periods in the lairage can lead to DFD if the animals are restless and fighting or mounting.

Animals should be as clean as possible at slaughter. Producers should wash their animals before leaving the farm. Trucks used for transport must be washed after each load and the lairage at the slaughterhouse should be kept clear of faecal matter and frequently washed.

### Stunning and Bleeding of Slaughter Animals

***Stunning Prior to Bleeding:*** Most countries have legislation requiring that animals are rendered unconscious (stunned) by a humane method prior to bleeding. Exceptions are made for religions which require that ritual slaughter without prior stunning is practised, provided the slaughter method is humane. Stunning also makes sticking (throat-slitting) less hazardous for the operator. The animal must be unconscious long enough for sticking to be carried out, and for brain death to result from the lack of blood supply.

### Methods of Stunning

*Direct blow* to skull using a club or poleaxe. The blow must be dealt with precision and force, so that the skull is immediately smashed, causing instantaneous unconsciousness. In cattle the aiming point is in the middle of the forehead in line with the ears, where the skull is thinnest. Horses have thinner skulls and are therefore easier to stun by this method. In sheep and goats the brain is more easily reached from the back of the neck. Pigs have a well-developed frontal cavity so the blow should be aimed slightly above the eyes.

*Slaughtering mask.* A bolt held in the correct position by the mask is driven into the animal's brain by a hammer blow. The device is usually fitted with a spring which returns the bolt to its original position.

*Free bullet* fired from a pistol into the skull is effective but unsafe. This method has been used on horses and cattle.

*Captive-bolt pistols* fitted with a blank cartridge are effective on cattle and sheep but not pigs whose skulls are thicker. After firing, the bolt returns to its original position in the pistol. The bolt may or may not be designed to penetrate the skull. With penetrating types the brain becomes contaminated with hair, dirt and bone fragments. If brains are to be saved as edible tissue then the non-penetrating type with a mushroom-shaped head should be used.

*Electrical stunning.* An electric current of high frequency but, in the case of manually operated equipment, of relatively low voltage (60–80 V) is passed through the brain of an animal for a few seconds to produce unconsciousness.

If applied correctly a deep state of unconsciusness is invariably achieved. Strict safety rules must be observed. Head tongs are suitable for pigs and sheep but not for cattle. The electrodes carried on the ends of the tongs must be accurately placed. Places where the skull is thick must be avoided. Electrical contact is impeded by hair and caked mud. Water or brine will improve contact but the head must not be completely wet otherwise the current will have a short-circuit path avoiding the brain. The electrodes must be applied with strong pressure.

*Carbon dioxide stunning* is used only in large pig abattoirs. Pigs are induced into a chamber and exposed to a concentration of 85 percent $CO_2$ for about 45 seconds. Although effective for anaesthetizing sheep, it is impractical because of large amounts of $CO_2$ collecting in the wool and affecting operators on the killing line.

### Bleeding After Stunning

The objectives of bleeding are to kill the animal with minimal damage to the carcass and to remove quickly as much blood as possible as blood is an ideal medium for the growth of bacteria.

Sticking, severing the major arteries of the neck, should immediately follow stunning. Care must be taken not to puncture the chest cavity or it will fill with blood.

*Cattle.* Insert the sticking knife carefully just above the breastbone at 45° pointed towards the head. Ensure that the carotid arteries and jugular veins are severed in one movement.

*Sheep.* Draw the knife across the jugular furrow close to the head severing both carotid arteries. Alternatively, the knife may be inserted through the side of the neck, though this requires more skill.

*Pigs.* As for cattle but do not go in too far or a pocket of blood will collect at the shoulder. To reduce contamination by the scalding tank water the cut should be as small as possible.

### Bleeding on a Rail

The most hygienic system of bleeding and dressing is to shackle the animal immediately after stunning, then hoist it on to a moving rail. The animal is stuck while being hoisted to minimize the delay after stunning. Bleeding continues until the blood flow is negligible when carcass dressing should begin without further delay.

Blood for human use must be collected with special equipment to avoid contamination from the wound, the gullet of the knife. A hollow knife directs blood away from the wound into a covered stainless-steel container without touching the skin or hide. The knife may be connected to a hose to reduce the risk of contamination.

The hose may even be connected to a pump to speed the blood flow. Between 40 and 60 percent of the total blood volume will be removed though this will be reduced if sticking is delayed. To prevent coagulation, citric acid solution made up with one part citric acid to two parts water is added at a rate up to 0.2 percent of the blood volume. The main sources of contamination during sticking and bleeding include the knife, the wound and the food-pipe. The knief should be changed after each operation and returned to a sterilizer. Cutting the hide of sheep and cattle and opening out to make a clean entry for the sticking knife reduces contamination from the wound. If the food-pipe is pierced semi-digested food may be regurgitated contaminating the blood and neck wound.

### Horizontal Bleeding

Horizontal bleeding is claimed to give faster bleeding rates and a greater recovery of blood. This may be due to certain organs and blood vessels being put under pressure when animals are hoisted, thus trapping blood and restricting the flow. Bleeding on the floor is very unhygienic. The operation should take place on a specially designed, easily cleaned stainless-steel table which should be cleaned frequently. If blood is to be saved it must not come in contact with the table before reaching the collecting vessel.

### Bleeding Without Stunning

The Jewish and Muslim religions forbid the consumption of meat which was killed by any method other than bleeding. Since it is difficult to guarantee that all animals will recover consciousness after being stunned by any particular method, stunning is not generally allowed. There are exceptions, however. Some communities do accept low-voltage electrical stunning. Because animals are fully conscious at the time of sticking, ritual slaughter may be less humane than sticking after stunning. To reduce the suffering operators must be highly skilled so that a successful gash cut severing all the veins and arteries is made quickly at the first attempt. Different communities have different regulations as to the orientation of the animal at sticking, some favouring a position lying on its side, others insisting it lie on its back. The animal should not be hoisted until unconsciousness due to lack of blood supply to the brain is complete.

### Scalding and Dehairing of Pigs (Using Simple Equipment)

Scalding in water at around 60°C for about six minutes loosens the hair in the follicle. Too low a temperature and the hair will not be loosened and too high a temperature and the skin will be cooked and the hair difficult to remove. The simplest equipment consists of a tank into which the pig is lowered by a hoist. The water is heated by oil, gas, electricity or an open steam-pipe.

To check the effectiveness of the scald, rub the skin with the thumb to see if hair comes away easily. Some machines have the thermostatic controls and timers. To reduce contamination, scalding water should be changed frequently, pigs should be as clean as possible at sticking, and bleeding should be fully completed before immersion.

In large factories pigs are transported through scalding tanks with rotating bars or through long scalding tanks stretching from the sticking point to the dehairing point in the time required for an effective scald.

Dehairing is done with a specially formed scraper (bell scraper or knife). If the scald is effective all the hair can be removed by this manual method. Another simple method is to dip the pig in a bath containing a hot resin adhesive.

The pig is removed from the bath and the resin allowed to set partially when it is peeled off pulling the hair with it from the root. This is less labour-intensive than scraping and produces a very clean skin. After use the adhesive is melted again, strained to remove the hair and returned to the tank.

Another method of removing dirt and hair in one operation is to skin the carcass though this is only done when the skin is required for leather goods.

With the simple scalding tank, dehairing and scalding may be combined in one operation. Inside the tank are rotating rubber-tipped paddles which are started after closing the lid. As the hair is loosened by the scalding water it is removed by the rubbing effect of the paddles against the skin.

Singeing removes any remaining hairs, shrinks and sets the skin, decreases the number of adhering micro-organisms and leaves an attractive clean appearance. It may be done with a hand-held gas torch. Automated systems transport the pig into a furnace and leave it long enough for an effective singe.

After singeing, black deposits and singed hairs are scraped off and the carcass is thoroughly cleaned before evisceration begins.

### Skinning of Cattle and Small Ruminants

***Cattle:*** The outer side of the hide must never touch the skinned surface of the carcass. Operators must not touch the skinned surface with the hand that was in contact with the skin.

*Combined horizontal/vertical methods:*

*Head.* After bleeding, while the animal is still hanging from the shackling chain, the horns are removed and the head is skinned. The head is detached by cutting through the neck muscles and the occipital joint. Hang the head on a hook. Lower the carcass on its back into the dressing cradle.

*Legs.* Skin and remove the legs at the carpal (foreleg) and tarsal (hind leg) joints. The forelegs should not be skinned or removed before the carcass is lowered on to the dressing cradle or the cut surfaces will be contaminated. The hooves may be left attached to the hide.

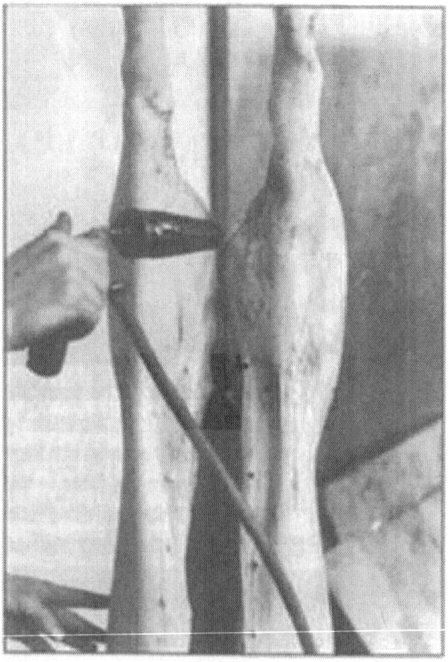

**Figure:** *After scraping away loose hairs any remaining hairs can be singed and the skin set with a hand-held gas torch*

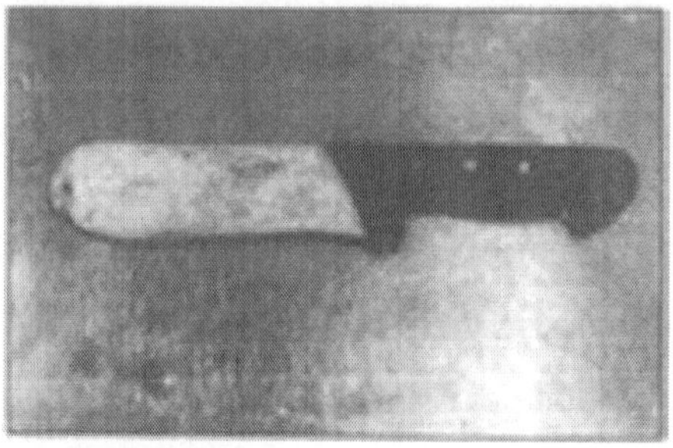

**Fiagure:** *A special knife, the black scrape, is used to scrape off any singed hairs and black deposits*

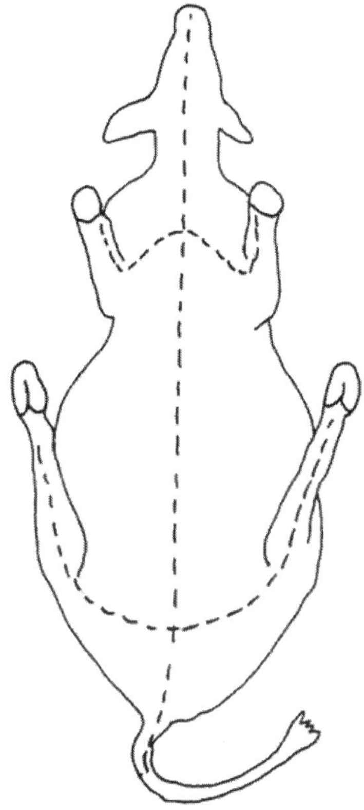

***Figure:*** *In the combined horizontal/ vertical dressing method the carcass is lowered on to a cradle, the legs, brisket and flanks are skinned, then the carcass is raised to the half-hoist position. Note that this is much less hygienic than verticaldressing on a rail*

*Flaying*. Cut the skin along the middle line from the sticking wound to the tail. Using long firm strokes and keeping the knife up to prevent knife cuts on the carcass, skin the brisket and flanks, working backwards towards the round. Skin udders without puncturing the glandular tissue and remove, leaving the supermammary glands intact and attached to the carcass. At this point raise the carcass to the half-hoist position, the shoulders resting on the cradle and the rump at a good working height.

Clear the skin carefully from around the vent (anus) avoiding puncturing it and cut the abdominal wall carefully around the rectum. Tie off with twine to seal it. Skin the tail avoiding contamination of the skinned surface with the hide. Raise the carcass free of the floor and finish flaying.

### Vertical Methods

High-throughput plants have overhead rails which convey the carcass from the sticking point to the chills. Hide removal is carried out on the hanging carcass. The operations are as in the combined horizontal/vertical method, but as it is not possible to reach the hide from ground level more than one operator is needed. A single operator may work with a hydraulic platform which is raised and lowered as required.

Automatic hide pullers are used in high-throughput slaughterhouses. Some types pull the hide down from the hind, others from the shoulders upwards towards the rump.

Automation of hide removal reduces contamination since there is less handling of the carcass and less use of knives. Moving overhead rails also improve hygiene by reducing carcass contact with operators, equipment such as dressing cradles and with each other since carcasses are evenly spaced.

#### *Small Ruminants*

Sheep fleeces can carry large volumes of dirt and faeces into the slaughterhouse. It is impossible to avoid contamination of sheep and lamb carcasses when the fleece is heavily soiled. The fleece or hair must never touch the skinned surface, neither must the operator touch the skinned surface with the hand that was in contact with the fleece.

### Combined Horizontal/vertical Method

The animal is turned on its back and cuts are made from the knuckles down the forelegs. The neck, cheeks and shoulders are skinned. The throat is opened up and the gullet (food-pipe) is tied off. The skin on the hind legs is cut from the knuckles down to the tail root. The legs are skinned and the sheep is hoisted by a gambrel inserted into the Achilles tendons. A rip is made down the midline and skinning proceeds over the flanks using special knives or the fists. The pelt is then pulled down over the backbone to the head. If the head is for human consumption it must be skinned or it will be contaminated with blood, dirt and hairs.

*Moving cratch and rail system*. The hanging carcass is lowered on to a horizontal conveyor made up of a series of horizontal steel plates, bowed slightly and divided into sets large enough to cradle a single animal. Two operators usually work together on each lamb performing the legging operations and opening the skin to the stage where it can be pulled off the back. When the gambrel is inserted into the hind legs it is hoisted on to a dressing rail.

### Vertical Method

At sticking the animal is shackled by one hind-leg and left to bleed. Dressing commences with the free leg which is skinned and the foot removed. A gambrel is inserted into this leg and hung on a runner on a dressing rail. The second leg is freed from the shackle, skinned and dressed, then hooked on to the other end of the gambrel. The skin is opened down the midline and cleared from the rump.

A spreader frame (a bar U-shaped at each end) spreads the front legs to simplify work on the neck, breast and flanks. The front toes are held in each end of the frame which is then slung up on to a separate travelling hook. The animal is therefore suspended by all four legs belly uppermost. Skinning continues as in the combined horizontal/vertical method.

To clear the shoulders and flanks, the forelegs are freed from the spreader and the feet removed, the animal returning to a vertical position. The skin can now be completely pulled off, including the head if this is for consumption, though this takes some work with the knife. In both methods, after fleece removal the vent and food-pipe are cleaned and tied off.

### *Evisceration*

With all species care must be taken in all operations not to puncture the viscera. All viscera must be identified with the carcass until the veterinary inspection has been passed. After inspection the viscera should be chilled on racks etc. for better air circulation.

### *Cattle*

The brisket is sawn down the middle. In the combined horizontal/vertical system this is done with the animal resting on the cradle. The carcass is then raised to the half-hoist position and when hide removal is complete the abdominal cavity is cut carefully along the middle line. The carcass is then fully hoisted to hang clear of the floor so that the viscera fall out under their own weight.

They are separated into thoracic viscera, paunch and intestines for inspection and cleaning. If any of the stomachs or intestines are to be saved for human consumption, ties are made at the oesophagus/

stomach, stomach/duodenum boundaries, the oesophagus and rectum having been tied off during hide removal. This prevents cross-contamination between the paunch and the intestines.

### Small Ruminants

A small cut is made in the abdominal cavity wall just above the brisket, and the fingers of the other hand are inserted to lift the body wall away from the viscera as the cut is continued to within about 5 cm of the cod fat or udder. The omentum is withdrawn, the rectum (tied off) loosened, and the viscera freed and taken out. The food-pipe (tied off) is pulled up through the diaphragm. The breastbone is split down the middle taking care not to puncture the thoracic organs which are then removed.

### Pigs

Loosen and tie off the rectum. Cut along the middle line through the skin and body wall from the crotch to the neck. Cut through the pelvis and remove the bladder and sexual organs. In males the foreskin must not be punctured as the contents are a serious source of contamination. All these organs are considered inedible.

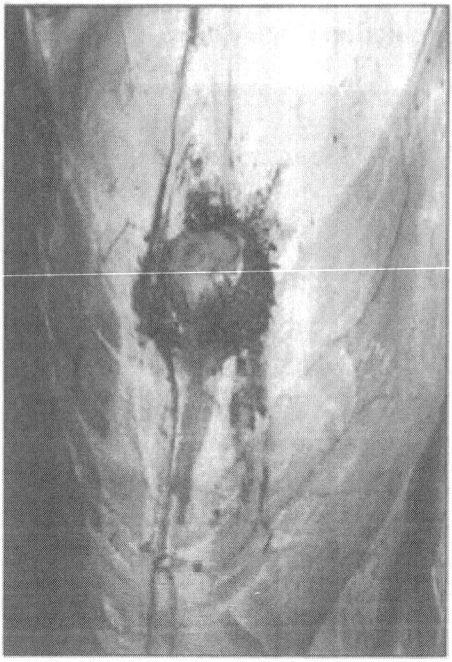

***Figure:*** *When cutting through the abdomen wall, if the viscera are punctured their contents will severely contaminate the carcass*

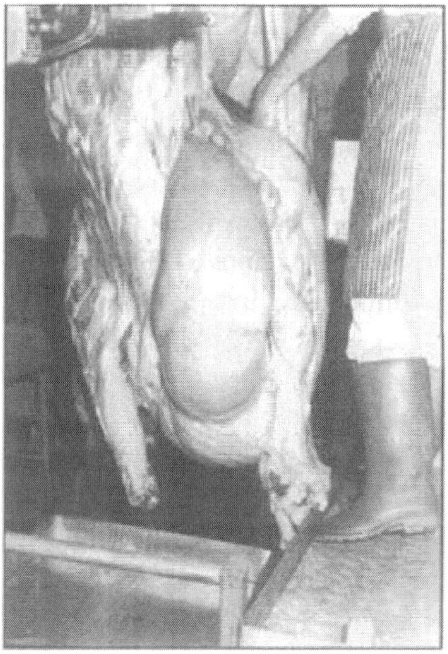

***Figure:*** *After carefully cutting the abdominal wall along the midline the viscera fall out under their own weight*

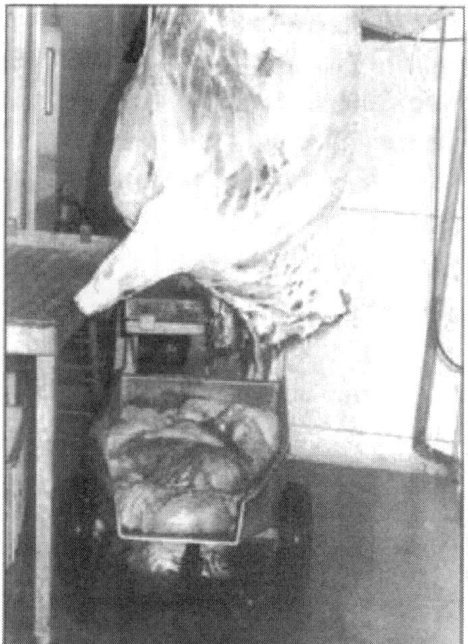

***Figure:*** *A suitable receptacle should catch the viscera so that they are not contaminated by contact with the floor*

***Figure:*** *A portable cart suitable for catching cattle stomachs and intestines with a separate tray for edible offal such as liver, heart and lungs*

Remove the abdominal and thoracic viscera intact. Avoid contact with the floor or standing platform.

The kidneys are usually removed after the carcass has been split down the backbone. The head is usually left on until after chilling.

### Splitting, Washing and Dressing of Carcasses

Hygienic carcass splitting with simple equipment

***Cattle:*** Work facing the back of the carcass. Split the carcass down the backbone (chine) with a saw or cleaver from the pelvis to the neck. Sawing gives a better result but bone dust must be removed. If a cleaver is used, it may be necessary to saw through the rump and loin in older animals.

The saw and cleaver should be sterilized in hot (82°C) water between carcasses. Power saws increase productivity.

### Pigs

These are suspended and are split down the backbone as for cattle, but the head is generally left intact.

### Sheep

Sheep and lamb carcasses are generally sold entire. If necessary they can be split by saw or cleaver, but a saw will probably be necessary for older animals.

### Carcass Washing

The primary object of carcass washing is to remove visible soiling and blood stains and to improve appearance after chilling. Washing is no substitute for good hygienic practices during slaughter and dressing since it is likely to spread bacteria rather than reduce total numbers. Stains of gut contents must be cut off. Wiping cloths must not be used.

Carcass spraying will remove visible dirt and blood stains. Water must be clean. Soiled carcasses should be sprayed immediately after dressing before the soiling material dries, thus minimizing the time for bacterial growth. Under factory conditions bacteria will double in number every 20 or 30 minutes.

In addition to removing stains from the skinned surface, particular attention should be paid to the internal surface, the sticking wound and the pelvic region.

A wet surface favours bacterial growth so only the minimum amount of water should be used and chilling should start immediately. If the cooler is well designed and operating efficiently the carcass surface will quickly dry out, inhibiting bacterial growth.

Bubbling of the subcutaneous fat is caused by spraying with water at excessively high pressure, which may be due to the pressure in the system or a result of holding the spray nozzle too close to the carcass.

### Carcass Dressing

The object of carcass dressing is to remove all damaged or contaminated parts and to standardize the presentation of carcasses prior to weighing. Specifications will differ in detail for different authorities. Veterinary inspection of carcasses and offal can only be carried out by qualified personnel.

Where signs of disease or damage are found the entire carcass and offal may be condemned and must not enter the food chain, but more often the veterinarian will require that certain parts, for instance those where abscesses are present, be removed and destroyed. Factory personnel must not remove any diseased parts until they have been seen by the inspector otherwise they may mask a general condition which should result in the whole carcass being condemned. Any instructions from the inspector to remove and destroy certain parts must be obeyed.

## Refrigeration, Handling and Transport of Carcasses and Meat

***Refrigeration of Carcasses:*** Carcasses should go into the cooler as soon as possible and should be as dry as possible. The object of refrigeration is to retard bacterial growth and extend the shelf-life. Chilling meat post-mortem from 40°C down to 0°C and keeping it cold will give a shelf-life of up to three weeks, provided high standards of hygiene were observed during slaughter and dressing.

Carcasses must be placed in the cooler immediately after weighing. They must hang on rails and never touch the floor. After several hours the outside of a carcass will feel cool to the touch, but the important temperature is that deep inside the carcass. This must be measured with a probe thermometre (not glass), and used as a guide to the efficiency of the cooling.

The rate of cooling at the deepest point will vary according to many factors including the efficiency of the cooler, the load, carcass size and fatness. As a general guide a deep muscle temperature of 6–7° C should be achieved in 28 to 36 hours for beef, 12 to 16 hours for pigs and 24 to 30 hours for sheep carcasses. Failure to bring down the internal temperature quickly will result in rapid multiplication of bacteria deep in the meat resulting in off-odours and bone-taint.

High air speeds are needed for rapid cooling but these will lead to increased weight losses due to evaporation unless the relative humidity (RH) is also high. However, if the air is near to saturation point (100 percent RH) then condensation will occur on the carcass surface, favouring mould and bacteria growth. A compromise between the two problems seems to be an RH of about 90 percent with an air speed of about 0.5 m/second. Condensation will also occur if warm carcasses are put in a cooler partially filled with cold carcasses.

The cooler should not be overloaded beyond the maximum load specified by the manufacturers and spaces should be left between carcasses for the cold air to circulate. Otherwise cooling will be inefficient and the carcass surface will remain wet, favouring rapid bacterial growth forming slime.

Once filled, a cooler should be closed and the door opened as little as possible to avoid sudden rises in temperature. When emptied, it should be thoroughly washed before refilling. Personnel handling carcasses during loading and unloading operations should follow the strictest rules regarding their personal hygiene and clothing and should handle carcasses as little as possible.

## Marketing of Meat Under Refrigeration

Chilled meat must be kept cold until it is sold or cooked. If the cold chain is broken, condensation forms and microbes grow rapidly. The same rules about not overloading, leaving space for air circulation, opening doors as little as possible and observing the highest hygiene standards when handling the meat apply.

An ideal storage temperature for fresh meat is just above its freezing point, which is about - 1°C (- 3°C for bacon because of the presence of salt). The expected storage life given by the International Institute of Refrigeration of various types of meat held at these temperatures is as follows:

| *Type of meat* | *Expected storage life at - 1°C* |
|---|---|
| ***Beef*** | ***up to 3 weeks (4–5 with strict hygiene)*** |
| Veal | 1–3 weeks |
| Lamb | 10–15 days |
| Pork | 1–2 weeks |
| Edible offal | 7 days |
| Rabbit | 5 days |
| Bacon | 4 weeks (at - 3°C) |

Under commercial conditions, meat temperatures are rarely kept at - 1°C to 0°C, so actual storage times are less than expected. The times would also be reduced if RH were greater than 90 percent.

Meat should be placed in the refrigerator immediately following receipt. Any parts which show signs of mould growth or bacterial slime should be trimmed off and destroyed. Hands must be thoroughly washed after handling such trimmings and knives must be sterilized in boiling water.

The refrigerator should be thoroughly cleaned after finding such meat and should also be cleaned on a regular basis.

Carcasses, quarters and large primals should not be cut into smaller portions before it is necessary as this will expose a greater surface area for bacteria to grow. Freshly cut surfaces are moist and provide a better medium for bacterial growth than the desiccated outer surfaces of cuts that have been stored for some time.

An accurate thermometre should be placed in the refrigerator and checked regularly. The temperature should remain within a narrow range (0° to + 1°C).

### Transport of Meat

Vehicles for transporting meat and carcasses should be considered as an extension of the refrigerated storage. The object must be to maintain the meat temperature at or near 0°C. Meat should be chilled to 0°C before loading. Meat should hang on rails, not on the floor. If stockinettes are put on carcasses they must be clean. Meat trucks should not carry anything other than meat.

The refrigeration is usually produced by injecting liquid nitrogen or carbon dioxide ($CO_2$) into the compartment or by blowing air over $CO_2$ chunks (dry ice). The temperature in these vans can be set and controlled to minimize the temperature rise and to avoid condensation on the meat surface. Insulated vans without refrigeration may be refrigerated by adding dry ice. While this is a reasonably good alternative to the refrigerated truck it does not allow the temperature to be controlled.

Uninsulated vans and open trucks should not be considered as suitable transport for meat, particularly in hot climates. In addition to the temperature abuse, condensation will occur when the meat goes back into refrigeration, and in open trucks the meat is exposed to attack from insects. Loading and unloading should be done quickly. If there are any unavoidable delays then dry-ice blocks should be placed in the partly filled van.

### Carcass and Meat Handling and Marketing Without Refrigeration

Where refrigeration is unavailable either owing to financial or technical reasons (e.g. no power supply), the shelf-life of meat is reduced to days or hours, not weeks. Slaughter and dressing must be near the point of sale and it must be quick and clean. If carcasses and meat are kept in well-insulated rooms, the temperature can be reduced with dry-ice blocks, if these are available. Since it is easier to chill boneless cuts rather than whole carcasses, hot-boning should be considered.

Stock must be handled carefully to avoid producing high-pH meat which will spoil more quickly. Rooms used for slaughter and handling meat must be clean and well ventilated, but out of direct sunlight, dust-free and verminfree (rodents and insects). Hot water (82°C) must be available to clean all equipment and surfaces and personnel must work very hygienically. Receive all blood into sealed containers and have separate skips on wheels for hooves, skins, green offal and trimmings.

Dressing on a vertical hoist will minimize contamination by floor or cradle contact. Let nothing drop on the floor, only into skips. Personal hygiene must be scrupulous. Any spills of gut contents on to the meat should be cut off, but careful work will avoid this. The dressed carcass should be hung on rails. If beef is quartered to facilitate handling, the cut surface is at risk.

Red offal should be hung on hooks. Any offal processing must be in rooms away from meat-handling facilities. Intestines for human consumption must be thoroughly cleaned and washed.

### *Storage and Transport Without Refrigeration*

Meat should be put on sale within a day of slaughter. If it has to be held it should be hung in a clean, well-lit hall with good ventilation. Insects, rodents and birds must be kept out, dust must not blow in. Trays of offal should be on shelves, not on the floor. Barrows for wheeling carcasses and quarters are better than carrying on shoulders, as they can be cleaned frequently. All staff must wear clean clothing and observe strict personal hygiene. Transport of non-refrigerated meat is very hazardous. If meat is to be put in stockinettes and sacks these must be very clean. Meat should be on rails in the truck or wagon, and it is not advisable to carry it more than a day's journey before sale.

## Meat Cutting and Utilization of Meat Cuts

### *Variations in The Sensoric Quality of Meat*

Large differences exist in the tenderness, juiciness and flavour of the various meat animal carcasses because of breeding, age, feeding and management. Within each animal carcasses and associated with the different muscles there are variations in tenderness that dictate how different cuts of meat should be prepared to yield the most palatable foods. Because of these differences in tenderness, juiciness and flavour, each meat cut should be merchandised according to its availability and palatability characteristics. Consequently, different prices should be charged for different cuts from the various meat animals so that consumers have choices. The tenderloin of beef is a relatively small cut and therefore of limited quantity but it is extremely tender and requires a minimum of cooking. Generally it is high-priced because of its high quality and consumer demand for a cut that is easy to prepare and serve. Roasts from the chuck or shoulder of beef are less tender than the tenderloin; however, when properly prepared by pot-roasting, they too will be tender, juicy, flavourful and will provide good nutritional value. Because there are more kilograms of chuck roast on any one beef carcass and because they require more time and effort to cook correctly, chuck roasts do not and should not demand the same high price per kilogram as tenderloin.

Throughout the world, countries have varied natural resources and capabilities for producing livestock and different methods must be used to utilize all meat products correctly and completely whether they are cut from cattle, goats, sheep, swine, deer or other animals and whether they come from the tender or less tender parts of those animals. In order to get the maximum eating satisfaction and also the maximum nutritional value, each cut must be matched with the correct cooking procedure. Loin cuts which are generally tender should be prepared by broiling or other dry-heat methods while cuts with considerable bone and connective tissue from the shanks should be either braised or simmered for stews and soups.

***Table:*** *Comparative differences in various compositional aspects of marketweight beef, pork and lamb*

| | ***Beef*** | ***Pork*** | ***Lamb*** |
|---|---|---|---|
| Average live animal weight (*kg*) | 454–544 | 95–104 | 45 |
| Age (*months*) | 36 | 6 | 8–12 |
| Dressing percentage (*carcass/live weight*) | 60 | 70 | 50 |
| Carcass weight (*kg*) | 272–318 | 68–73 | 23 |
| Carcass composition (%) | | | |
| Lean | 52 | 50 | 55 |
| Fat | 32 | 32 | 28 |
| Bone | 16 | 18 | 17 |

Generally, meat animals should be maintained in an environment that permits optimum growth and development. Animals gaining weight rapidly are usually in good condition and the meat derived from their carcasses will be fatter, juicier and richer in flavour. Additionally, the amount of meat in proportion to hide, bone and offal will be greater.

The age to slaughter animals varies depending on many things. The highest quality beef comes from animals that are under 36 months of age. Old cows produce highly acceptable beef if properly fattened and processed. Depending on the calf and the feeding regime, calves are best slaughtered between three and 16 weeks of age. Hogs may be killed any time after they reach six weeks of age, but for the most profitable pork production may need to be fed for five to ten months. Sheep and goats may be killed anytime after six weeks, but the more desirable age is from six to 12 months.

All meat animal carcasses are composed of muscle, fat, bone and connective tissue. The chief edible and nutritive portion is the muscle or lean meat. The muscle is seldom consumed without some of the attached fat and connective tissue. The carcass composition of animals slaughtered after usual fattening periods is shown in table. It can be noted that the carcass composition varies little between species and is some what dependent on the fatness of the animal at slaughter.

The lean of each meat animal carcass consists of about 300 individual and different muscles of which only about 25 can be separated out and utilized as single muscle or muscle combinations. The separated muscles are not all the same. They vary widely in palatability (tenderness, juiciness, flavour) depending on the maturity or age of the animal and the body location from which they were taken.

Generally, muscles of locomotion found in the extremities or legs are less tender and more flavourful than muscles that simply support the animal such as those found along the back. The latter are usually more tender and less flavourful. Other factors may influence palatability but maturity and body location are probably the most important.

Colours of the lean and fat are important characteristics of a normal, wholesome products. Most diseased or unnatural conditions will change the colour from what is considered normal for the species. Generally the colour of the fat will be from pure white to a creamy yellow for all animals. Pink or reddish fat probably means that the animal had a fever or was extremely excited prior to slaughter. The colour of the muscle tissues for normal product should be:

| ***Meat*** | ***Colour*** |
|---|---|
| Beef | Bright cherry red |
| Goat meat | Light pink to red |
| Lamb | Light pink to red |
| Pork | Greyish pink |
| Veal | Light pink to red |
| Venison | Dark red |

Almost always tissues from older animals are darker in colour. At times the fat on some carcasses from young animals will be dark yellow because of the breed which lacks the ability to convert yellow carotene to colourless vitamin A and/or because the animals have consumed large amounts of green forage. It is not uncommon for aged ruminant animals to have carcasses with yellow fat.

At times animals will suffer from stress prior to slaughter and signs of their reaction will be evident in the carcass. Stressed cattle often produce dark cutters in which the muscle is not the normal bright cherry red but rather is dark red and sticky. Hogs suffering from porcine stress syndrome (PSS) prior to slaughter may yield carcasses that are pale, soft and exudative (PSE) or dark, firm and dry (DFD). Exudative carcasses are watery and rapidly lose water. None of these conditions produced by ante-mortem stress renders the product inedible but both lower the palatability and eye appeal of the beef and pork and can be confused with other more serious disease conditions.

### Equipment For The Meat-cutting Operation

- solid cutting table, preferably made of non-corrosive material (stainless steel, aluminium or galvanized material) with hard plastic top. If wood has to be used instead of plastic only tight wooden tops/cutters should be used.
- oil or water sharpening stone
- sharpening steel
- knives
    - o boning - 20 cm straight
    - o steak - 30 cm curved
- meat saw - hand or electric
- totes, bins and meat trucks (plastic or other non-corrosive material)
- wrapping table
- paper or plastic foil/bags for meat wrapping
- tool holder
- metal mask/safety gloves
- boning aprons/safety aprons
- hand wash-basin
- knife sterilizer

### Beef Cutting

Four essential points when cutting beef (or any other meat animal carcass) are:

- Cut across the grain of meat when possible.
- Use sharp knives and saws for speed and good workmanship.
- Keep the cutting table orderly and have a place for everything.
- Be clean and sanitary in all operations.

There are different ways to cut the fore- and hindquarters of beef depending on its use, the wishes of the consumers, and the quality of the carcass. Poor-quality meat is normally used for further processing, while higher-quality and thicker-fleshed carcasses are used as fresh meat in the form of steaks and roasts.

### Halving

Halving is done immediately after the animal has been dressed and every effort should be made to saw the carcass into equal sides through the centre of the backbone.

### Quartering

Quartering or ribbing down is the division of a side of beef between the twelfth and thirteenth ribs into fore-and hindquarters. One rib is usually left on the hindquarter to hold the shape of the loin and to make it easier to cut steaks.

Dividing between the twelfth and thirteenth ribs splits the carcass almost in quarters, usually with slightly heavier forequarters. Make this cut straight and neat. Locate the exact place between the ribs on the inside of the carcass and make the cut about 5 cm from the midline at the flank. The flank part should be left attached until the quarter is ready to be carried to the cutting table. Then saw the backbone, making the cut even with the incision that was made with the knife to produce a smooth and attractive appearances to the small end of the loin. Make this cut from the inside. The large muscle exposed when this cut is made is the "eye of beef" in which most of the quality characteristics of the meat can be seen including colour, marbling, firmness and texture. High-quality beef will have a bright cherry-red colour, some intramuscular fat or marbling, be firm to the touch and fine in texture.

When the person carrying the meat has a firm grip on the forequarter, the small strip of flesh holding the quarters together should be cut. With some practice and experience, one can learn to carry a forequarter easily by holding below the shank so that the full weight of the quarter is on the carrier's shoulder when it is cut down. By taking a step forward as the cut is being made, it is easier to have the quarter drop with the right proportion of weight on the shoulder. The right forequarter should be carried on the left shoulder and the left forequarter on the right shoulder. When placing the forequarter on the cutting table, always have the inside up.

### *Bone-in method*

By far the easiest way to merchandise meat is to have some basic information relative to the bone and muscle structure of the carcass and to utilize an electric saw to cut up the whole carcass. This is now being done to a large extent by meat packers who cut out what is commonly referred to as a wholesale or primal cut such as a whole chuck (shoulder), rib, loin or round of beef. The cut may or may not be trimmed of some bone and fat and then vacuum-packaged and shipped to a retail store. The vacuum-packaging provides an anaerobic atmosphere and the refrigerated shelf-life of the product may be extended as much as two or three months. The store personnel need have only the slightest knowledge of meat cutting. The primal is positioned correctly and run across the saw in a prescribed fashion, the saw dust is scraped off, and the consumer-sized cut packaged for retail sale.

Common wholesale or primal cuts of beef from the forequarter are the square-cut chuck, shank, brisket, plate and rib, and from the hindquarter the flank, loin and round. The kidney knob consisting of kidney and fat is removed from the loin. Since the hindquarter contains a higher proportion of tender cuts, it is usually in greater demand and returns higher prices.

*Forequarter*. The first cut to make is between the fifth and sixth ribs counting from the neck back . This cut is made parallel with the ribs and produces a cross-cut chuck consisting of a square-cut chuck (also called chuck and blade), foreshank and brisket. Next the foreshank and brisket are removed by cutting through the first sternal cartilage (the first soft segment of the breastbone), and making the cut almost parallel with the backbone of the carcass.

*Foreshank.* The foreshank is separated from the brisket by following the natural connective tissue seam between the muscles with a knife. The foreshank can then be sawn into small pieces to be used for soup stock or the lean may be removed and used for ground meat.

*Brisket.* The brisket, boned and made into a roll, can be used either as a pot roast or can be cured (corned).

*Square-cut chuck.* This wholesale cut contains the first five ribs of the forequarter and may be sawn into steaks or roasts. Several cuts are usually made across the bottom or shank end of the chuck resulting in arm steaks or roasts. The chuck is then turned and cuts are made parallel with the ribs, resulting in blade steaks and roasts. If the carcass is of high quality and thickly fleshed, steaks cut from the rib end of the chuck or across the arm bone will be highly desirable. Blade cuts to be used as roasts should contain two or three ribs and should be trimmed as for standing rib roasts, although for convenience in carving all bones may be removed. The portions nearest the neck usually have more connective tissue and are recommended for simmering rather than for steaks and roasts.

Only the neck remains to be processed. It is usually severed at a point where it enlarges to meet the shoulder. The neck contains a large amount of bone and connective tissue and is generally used for simmering, corning or grinding. All bloody portions should be trimmed off before other cutting is done.

*Short plate.* The cut to divide the short plate from the rib is made 18–25 cm from the inside edge of and parallel with the chine or backbone. This division varies according to the thickness of the carcass. With a thick carcass, the cut may be made further down the ribs, and with a thin carcass nearer the spinal column.

The plate may be used for different purposes, but it is commonly used for stews or further processing. Short ribs, which are suited for broiling, are also cut from the upper portion of the plate, usually about 5–8 cm in length. If the plate is to be used for corning, all of the ribs should be removed. If used for stews, the ribs can be left in and the plate sawn crosswise into small pieces. The plate can also be boned and the meat used for ground meat or sausage products. Before cutting the plate in any way, remove the tough membrane lining the inner portion below where the ribs join the breastbone.

*Rib.* The rib cut is made up of the rear seven ribs in the forequarter. This is the most valuable piece of meat from the forequarter because it is the most tender and has the least amount of bone. It has a large bundle of muscle fibre that runs parallel to the backbone.

There are several different ways to prepare the rib cut for cooking as a roast. It may also be used for steaks. It may be prepared as a bonein, folded or rolled roast. If prepared as a bone-in roast, the superior spinous processes of the vertebrae or featherbones are loosened from the meat and then cut off with a saw.

In making this cut, keep the knife as close to the bone as possible to avoid removing the thin lining that surrounds the bundle of muscle fibre next to the bone. With the saw, cut across the ribs at intervals of about 8 cm, just deep enough to cut through the ribs. Also remove the yellow connective tissue or ligament found between the outer covering and the layer of muscle.

The only difference between bone-in and a folded rib roast is that a small 5-cm piece of rib is removed so that the thin end of the cut may be folded and skewered to the heavy portion. This simply makes a neater, more compact package.

*Hindquarter.* Place the hindquarter on the cutting table with the inside of the carcass up because the first cut made is to remove the kidney knob from the inside of the loin. (However, loosening of meat cuts is also possible from the hanging beef side or beef quarter.)

*Kidney knob.* Begin removing the kidney fat at the lower end and loosen it with a knife where it is attached to the loin, leaving a thin covering on the inside of the loin and being careful not to cut into the tenderloin muscle.

*Flank.* Remove the flank next by cutting into the scrotum or udder, following the round muscle and cutting close enough so little of the lean meat is taken from in front of the stifle joint. Continue cutting along and below the outer portion of the line of the kidney fat, or in a straight line to leave 10 cm of the thirteenth rib in the flank. This cut may vary with the thickness of the carcass and is lowest in thick or heavy carcasses.

The tough membrane covering the inside of the flank must be removed by cutting off a thin strip on the lower side and then peeling off the membrane. A small piece of lean meat on the inside of the end portion of the flank, weighing 1.2–1.4 kg, is known as the flank steak. This heavy bundle of muscle fibres is dry and if used for steak is often scored on both sides, marinated or sliced thin to make it more tender and desirable as a steak. The entire defatted flank may be used for stew or ground beef or rolled around stuffing and pot-roasted.

*Round.* The round and loin are divided at about the fourth sacral joint in the spinal column to almost parallel with the back end of the round, or to about 5 cm in front of the stifle joint. The aim is to cut the tip of the ball-and-socket bone in the hip joint, cutting off a piece about 2.5 cm in diametre. The round includes the rump, round cushion (consisting of knuckle piece and inside round muscle or topside), outside round muscle (also called bottom round muscle or silverside) and hind shank.

Remove the rump by cutting just below the exposed pelvic or aitchbone. The rump usually has a large amount of bone. The most desirable piece of rump is cut from the upper portion and is composed of eye and bottom round muscles. The removal of bone and tying the rump means that it requires less oven space and is easier to carve.

Round steak is cut in comparatively thin slices from the full round after removal of the rump. The choicest round steaks are cut from the centre section.

The remaining portion is made up of the hind shank and the piece called the heel of the round. The heel of round is used as a pot roast and is removed by cutting close to the bone and tearing away as much meat as possible from the backside. The shank can be sawn into pieces to be used for soup stock.

*Loin.* The loin is usually completely sawn into steaks beginning at the large end. Sirloin steaks are cut first and the first three or four are known as wedge or round bone sirloin steaks. These are the least desirable pieces of the sirloin. The last sirloin is cut where the hip-bone is separated from the spinal column and the steak cut there is known as the hip-or pin-bone sirloin steak.

The small portion of the loin known as the short loin is the source of Tbone steaks. This area contains the two most tender muscles in the whole carcass, namely, the loin eye muscle above the bone and the tenderloin muscle below the bone. T-bone steaks are cut to about 10 cm from the end of the short loin. This tip portion can either be used as a roast or be cut into rib steaks. Rib steak from the short loin is identified by the piece of the thirteenth rib remaining on it.

When beef is to be cured and dried, pieces should be taken from either the chuck or the round. If the round is used, remove the rump and follow the procedure for muscle boning. If taken from the chuck, use the heavy muscle lying over the outside of the shoulder-blade commonly known as shoulder clod.

### Muscle-boning Method

One excellent approach to the cutting up of meat animal carcasses which is becoming more popular and utilized by large meat processors is the procedure commonly referred to as "muscle-boning". While this procedure is particularly adaptable to large carcasses such as beef, it can be successfully used on carcasses or cuts of any size. Muscle-boning is also popular among hunters who do not have meat saws but who want to cut up a whole carcass with a knife while removing the bone that would otherwise fill valuable freezer space. Any animal carcass with a complete and thick layer of subcutaneous or cover fat would have to have most of the fat removed in order to expose the muscles.

Once the fat is removed, a boning knife can be used to separate each large individual muscle or group of muscles. This is done along the seams of connective tissue that encases each muscle. Once separated the muscle mass is then cut from the bone, thus the term "muscle-boning". The advantages of this procedure are numerous; however, the principal reasons for using it are to obtain small-sized portions for sale or preparation; to permit each muscle or muscle combination to be treated or prepared according to its individual characteristics of size, tenderness, flavour or fibre orientation; and to remove much of the bone and fat that would otherwise take up packaging and storage space.

Directions for muscle-boning a side of beef are given here. Initially for muscle-boning, the side of beef is divided into fore-and hindquarters as described for the bone-in method. Also, both the fore-and hindquarters are placed on the cutting table with the inside up. One muscle-boning method is as follows:

*Forequarter.* The forequarter is sawn into square-cut chuck, foreshank, brisket, rib and plate as in the bone-in method.

*Foreshank.* The foreshank has attached to it, behind the elbow joint, a relatively large, thick piece of muscle. This is usually cut out by following the connective tissue seams and produces a fairly large triangular-shaped cut correctly identified as boneless arm roast. The remainder of the foreshank can be sawn into soup bones or can be separated into bone and soft tissue with a knife. The soft tissue is composed of muscle, fat and a large amount of connective tissue which is best utilized as ground meat.

*Brisket.* The ribs and sternum are lifted from the inside of the brisket and the excess fat is removed. The brisket can either be rolled and tied to be used as a pot roast or it can be cured.

*Square-cut chuck.* The neck is sawn from the chuck and trimmed of bone, fat and the large prescapular lymph gland. The boneless neck can be utilized as a pot roast; however, it is more often cut into cubes for stew or ground meat.

From the large remaining portion of the chuck, the ribs and feather bones (superior spinous processes) are removed with a knife and the heavy, yellow connective tissue or elastin is removed from the top of the cut. With a knife the thick portion is then separated into outside and inside portions by following the inside or smooth side of the blade-bone which is then lifted from the outside piece along with what remains of the arm bone. The inside portion which contains some of the rib eye muscle is often rolled and tied to be used as a pot roast. There is a part of the outside chuck, a muscle that somewhat resembles the tenderloin muscle in size and shape but not in tenderness, which is often cut into steaks known as chuck fillets.

*Rib.* The rib is prepared by first sawing across the rib bones to facilitate the removal of both the backbone and the ribs with the knife. Another procedure often used to bone out a rib is carefully with a sharp knife to loosen the small strip of meat found between the ribs. The ribs are then loosened by cutting close to the bone and removed by striking with a blunt instrument. After removing all bones and the heavy yellow connective tissue, the meat may be rolled into a tight bundle with the thin portion on the outside and tied tightly. Preparing ribs in this way makes for convenient carving and requires less cooking and storage space. About 25 percent of the initial rib weight is lost when the bones are removed. The boneless rib may also be sliced into boneless rib steaks.

*Plate.* After the heavy connective tissue lining is peeled from the inside of the plate, the bones are removed and the lean meat cubed for stew or prepared for grinding in a way similar to the trimming of the brisket.

*Hindquarter.* As a first step, the kidney and accompanying fat are removed from the hindquarter carefully with a knife so as not to cut into the tenderloin muscle. The hindquarter is then separated into flank, round and loin as described in the bone-in method.

*Flank.* Remove the flank by cutting into the scrotum or udder, following the round muscle and cutting close enough so that little lean meat is taken from the front of the stifle joint. Continue cutting along and below the outer portion of the line of the kidney fat in a straight line and saw through the thirteenth rib. Again the flank steak is removed as described in the bone-in method.

*Round.* The round and loin are separated with a saw as described in the bone-in method. The pelvic bone is removed from the round and the muscle sections of the round are exposed.

Muscle-boning the round means that the large muscle masses of the round are separated from each other by following the natural connective tissue seams. In front of the stifle joint, the tip or knuckle piece is removed, then the topside or inside round muscle, and then the remaining silverside or bottom round muscles.

The latter is often divided and the eye of the round removed separately. All of the separated muscles may then be used as roasts or sliced into steaks. Muscle-boning is particularly useful when beef is prepared for roasting for large groups such as pit barbecuing.

*Hind shank.* The hind shank, somewhat like the foreshank, has a large muscle group attached to it that can be removed and utilized as a pot roast. This cut is sometimes referred to as the "duck" of beef.

*Loin.* The tenderloin muscle is carefully cut from the inside of the loin and usually cut into individual steaks. The remainder of the loin is then sawn just in front of the hip-bone into the short loin and sirloin sections. The bone is removed from the sirloin which is a somewhat complicated procedure because the pelvic bone is fused with the backbone. The short loin is boned and the muscle that is known as boneless top loin is usually cut into boneless top loin steaks.

### On-the-rail Boning

This is a modification of the muscle-boning method. Typical for on-the-rail boning is the hanging position of the hindquarter or the entire beef side during the boning procedure.

The removal of the different meat cuts from the hanging carcass is considerably facilitated. Beef cuts can easily be pulled downwards under their own weight after cutting them free along their natural connective tissue seams. Special hooks with handles used by the operators are an additional aid for the correct fixation of the cuts during boning.

On-the-rail boning is the most hygienic way of meat cutting. Contamination by hands of operators, tools, cutting-boards, etc. is less than with other methods.

The technique is also suitable for smaller operations. Final trimming of the meat cuts takes place on cutting tables as usual.

When meat cuts are produced by muscle-boning it is often difficult to identify them, primarily because traditionally the size and shape of the accompanying bone has been used as the major means of identification. Also, the traditional shape of muscle in a cut of meat is often determined because of its attachment to bone. Many conventional cuts of meat combine muscles because of their association, size and proximity to bone or general location. The basic principle of merchandising meat is to separate the tender from the less tender and to sell each according to its palatability characteristics and its possible method of preparation. Muscle-boning facilitates this type of merchandising.

### Pork Cutting

Halving is done immediately after the animal has been dressed and every effort should be made to saw the carcass into equal sides through the centre of the backbone. The side to be cut should be laid on the cutting table with the inside up.

The primal cuts of pork are: ham, fore-end or forequarter, loin and belly.

*Hind foot.* The hind foot is removed by sawing through the hock joint at a right angle to the long axis of the leg.

*Ham.* The ham may be removed in several ways to make either long-cut or short-cut hams. One procedure (short-cut) is to locate the division between the second and third (or the third and fourth) sacral vertebrae and saw perpendicularly to the long axis of the ham. After the bone has been severed with the saw, the knife is used to complete the removal of the ham. The ham is further trimmed by removal of the tail bone on one side and the flank on the other side. Commonly a skinned ham is produced by removal of three-fourths of the skin and fat from the rump end. For the production of special cured dried hams the skin is left on.

### The Pork Carcass and its Bones

In order to obtain a long-cut ham the division is made between the last two (fifth and sixth) lumbar vertebrae. The long cut is composed of a rump or chump portion and a leg portion comprising centre section and shank portion. Nowadays more processors are removing the bones thus fabricating a boneless rump (chump) and a boneless ham. The ham is commonly merchandised in smaller portions (topside, silverside, thick flank, shank).

### The Pork Carcass and its Cuts

The cutting procedure of the ham is as follows. Remove tail bone and aitch bone and cut the rump off. Peel back the rind and associated fat to expose the topside muscle on the interior side of the leg. Separate the topside by following the natural seam between it and the silverside (outside portion of leg) and thick flank (front position of leg). The topside can then be sliced into steaks. This produces between five and six lean steaks depending on the thickness and weight required by the customer. The next step is to remove the leg bone (femur). The thick flank (knuckle) is cut from the silverside by following the natural seam. Remove the kneecap (patella) and the internal fat deposits before further preparation of the thick flank, e.g. for diced pork or steaks.

*Forefoot*. The forefoot is removed by sawing through the junction between the foreshank and the forefoot bone at a right angle to the length of the foot. This foot contains some muscle and is therefore more desirable than the hind foot for food.

*Fore-end*. Considerable variation exists as to where the fore-end is removed. Generally one to three ribs are left on the pork fore-end. Locate the division between the third and fourth ribs from the head end and saw perpendicularly to the length of the backbone. The fore-end is trimmed of the hock which is cut off about halfway up the leg and about two-thirds of the skin and fat is removed from the butt or top end. Additionally the neckbone (all cervical and three thoracic vertebrae) and the jowl or cheek meat are removed. The jowl is removed by a straight cut parallel to the cut that separates the fore-end from the side just behind the site where the ear was removed. The fore-end may be divided into two cuts (spare-rib, also called blade Boston, and hand, also called arm picnic) by sawing just below the exposed lower end of the blade-bone parallel to the top of the shoulder. The spare-rib can be sliced into steaks or used as a roast. It can easily be made into a boneless cut by removing the corner of the blade-bone.

Besides this method some other ways of cutting and boning the pork foreend exist. In order to obtain boneless cuts (shoulder and neck-end) from the fore-end the following technique is recommended. Seam the shoulder carefully from the rest of the side, leaving the rind and associated fat behind. Release the under-blade steak and remove the blade-bone (scapula) and the shoulder-bone (humerus). Separate the main muscle block from the smaller group. The smaller group, after trimming the fat off, can be used for dicing. The main shoulder block should be trimmed of excessive connective tissue. It can be separated further into the blade and feather muscles and the main shoulder muscle. These can then be sliced into a number of boneless steaks. The group of muscles on either side of the spinous processes of the neckbone and the two or three following segments of the backbone is called the neck-end. The neck-end is loosened from the backbone and after trimming off excessive rind, fat and any adhering ragged edges it can be cut into attractive steaks.

*Lion.* The middle or centre section of the pork side is divided into loin and belly by a straight cut from the edge of the tenderloin muscle on the ham end through a point on the front rib tight against the protruding edge of the split backbone. The fat back (skin and excess fat) is removed from the loin so that a complete fat cover about 0.5 cm thick remains. Starting along the backbone side at the shoulder end, cut and lift the fat over the curve of the loin muscles without cutting into the lean. The loin can be roasted whole, cut into smaller roasts or cut into chops. Shoulder, rib, loin and sirloin chops are made from the loin. Chops for broiling or frying should be cut 1.3–1.9 cm thick. Thicker chops may be made and a pocket cut into them for stuffing.

*Belly.* Separate the spare-ribs from the belly by cutting closely underneath the ribs beginning at the flank end. Prepare the bacon side from the belly by removing any thin or ragged pieces of lean. Turn the belly over and remove the lower edge with a straight cut just inside of the teat line. Trim the flank edge of the belly to square the whole piece to prepare it for curing.

### Lamb Cutting

***Method:*** This procedure as described may also be followed for the processing of deer, goats, sheep or other animal carcasses of similar size.

***Cooling:*** All lamb carcasses should be promptly chilled and kept at a low temperature (-2° to 2°C) until cut and utilized. Do not permit lamb carcasses to freeze within a day after slaughter or the meat may toughen. Lamb carcasses can be cut into retail cuts after they have been chilled for 24 to 48 hours.

***Carcass:*** Lamb carcasses are generally not split into halves after dressing because they are not thick enough in any location to create cooling problems. Begin cutting the lamb carcass by removing the thin cuts, i.e. flank, breast and foreleg. Lay the carcass on the cutting table and mark one side from the cod or udder fat in front of the hind leg to the elbow joint. After removing the thin cuts from both sides, remove the kidneys, kidney fat and diaphragm. Next the carcass is turned over and the neck removed either in thin slices to be braised or in one piece to be added to stew or to be boned and ground.

The trimmed carcass can then be separated into four primal cuts, each with different characteristics. A cut between the fifth and sixth rib removes the shoulder. Another cut between the twelfth and thirteenth (last) rib separates the rib from the loin. The loin and legs are separated just in front of the hip bones by cutting through the back where the curve of the leg muscles blends into the loin.

### *The Lamb Carcass and its Bones*

*Legs.* Split the legs through the centre of the backbone. Trim off the flank and cod or udder fat. Utilize the saw and knife to remove the backbone from the leg. The leg may be further trimmed by cutting through the knee-joint which is located about halfway between where the muscles of the shank end and the muscles of the lower leg begin. Work the knife and cut through the joint. Several sirloin chops may be cut from the loin end of the leg. Legs may either be prepared with the bone in or the bones completely removed and the leg rolled and tied.

*Loin.* The loin is usually split through the middle of the backbone and chops are cut perpendicularly to the backbone. Lamb chops are cut about 2.5 cm thick. Double or "English" chops are made from a loin that has not been split. Remove the fell or connective tissue covering before cooking chops.

*Rib.* The rib of lamb is prepared by sawing through the ribs on both sides of the backbone. The main portion of the backbone is then removed with a knife. Rib chops are easily made by cutting between the ribs. Remove the fell before cooking the chops. The breast portion may be barbecued in one piece or made into riblets by cutting between the ribs.

*Shoulder.* After splitting through the backbone, the shoulder may be roasted as is, made into chops, or boned and rolled into a roast. Arm chops should be made first by cutting parallel to the surface where the foreleg and breast were removed. Blade chops are made by cutting between ribs and sawing through the blade- and backbones. To prepare a boneless shoulder, first remove the ribs and backbone by cutting closely underneath the ribs, backbone and neck vertebrae. Next from the rear surface cut along the inside of the blade-bone to expose it and the armbone. Cut along the edges of the bones and remove them. Roll the meat and tie it securely with clean twine. The boneless shoulder may also be made into a pocket roast and stuffed with ground lamb or other dressing. The edges of the pocket roast are stitched together.

*Shanks.* Both the fore- and hind shanks when removed can be barbecued, cut into pieces for stew or boned and the meat ground.

*Lean trimmings.* Lean trimmings of lamb in chunks are suitable for stews or to be marinated and used for special roasts. Other lean trimmings can be ground and used as one would prepare ground veal or beef.

### Hygiene Rules For Marketing Chilled Meat Cuts

Chilled meat is usually kept for the sale in refrigerated display cabinets, either unwrapped or portioned and packaged for self-service outlets. Refrigerated display cabinets may have fan-assisted convection and/or natural convection. Fan-assisted types are better able to maintain a lower temperature as they are less affected by draughts. Cabinets should be stacked to maintain a good air flow around all meat.

Do not store or display unwrapped cooked and raw meat together. Use separate refrigerators, display cabinets etc. to avoid cross-contamination. Raw-meat exudate on to cooked meat gives an explosive bacterial growth.

Simple packaging of fresh meat with plastic foil has become very popular with the availability of suitable and inexpensive film. The main objective of simple packaging is to provide hygienically protected portioned meat for self-service retail outlets. But the meat portions must also satisfy the customers' preference for bright red fresh meat. This colour is due to the pigment myoglobin loosely binding oxygen to form oxymyoglobin. For this colour to develop and be maintained, the wrapping film must have a high-oxygen permeability. To avoid desiccation of the cut surface, the film should have a low-moisture permeability. After a time the cut surface becomes more brown as a result of myoglobin binding the oxygen more tightly to form metmyoglobin. This may take up to three days depending on the temperature, the number of bacteria and other conditions.

Simple packaging for retail sale in self-service outlets usually involves placing the meat portion in a plastic tray and overwrapping with a clear plastic film. Plastic trays are more hygienic than cardboard. The portions cut should be based on local demand and only a day's sales should be cut at a time.

The principal object of this type of simple packaging from a hygiene point of view is to reduce contamination from airborne micro-organisms. High standards of hygiene are required in the cutting and packaging operations. On large pieces of meat the bacteria mainly colonize the outer surfaces. When meat is cut even with a clean knife they will be spread on to the freshly cut moist surface and multiply rapidly. This is not an argument for relaxing hygiene standards, rather it underlines the need not to add to the bacterial load by further contamination.

All surfaces and tools in the cutting and packaging room must be kept thoroughly clean. Packaging materials should be stored in hygienic conditions protected from dust and attack from insects or vermin. It is most important that personnel involved in cutting and packaging pay particular attention to personal hygiene as they are the most likely source of food-poisoning pathogens which may survive better in the package environment than on unpackaged meat. This is in part due to the packaging preventing surface desiccation. The moist surface favours bacterial growth as does the high relative humidity that builds up within the pack.It is important to retard bacterial growth by maintaining a low temperature during the display life of the packs. Overwrapping actually increases the meat temperature as the layer of trapped air acts as an insulator. Heat generated by light warms the upper surface. Meat should be thoroughly cooled before packaging to help maintain a low temperature during its display life.

Mincing meat spreads bacteria on the surface all through the meat which therefore has a shorter shelf-life than cuts. Mince may be packaged and overwrapped but the mincer must be kept scrupulously clean and the packs kept well chilled. Only small quantities of mince should be prepared at a time.

Cooked meats, which typically have much lower bacteria counts than fresh, are more open to attack from airborne micro-organisms as these will be faced with little competition. Packaging is therefore particularly beneficial in preventing this type of contamination for cooked meats.

Bacteria introduced during cutting and packaging face little competition and may be of the food-poisoning type if personal hygiene is poor. If very high standards of hygiene cannot be maintained then a pasteurizing treatment after packaging will be necessary. Even this, however, will not guarantee destroying *Bacillus* and *Clostridium* spp. if these have been introduced.

### Cooking Methods For Different Meat Cuts

Primarily because of natural tenderness or lack of tenderness, different cooking procedures are utilized to prepare the various cuts of meat correctly. Tender cuts are best cooked with dry heat, as by broiling, roasting or pan broiling. Less tender cuts are tenderized by cooking with moist heat. Connective tissue is softened and made tender by cooking slowly in moisture.

Temperature control is important in meat cookery. Meat loses moisture, fat and other substances such as soluble proteins during cooking. Cooking losses can be minimized by controlling the cooking temperature and the final internal temperature of the meat. Higher oven and higher internal temperatures increases shrinkage. Whenever possible a meat thermometre should be used to determine accurately the degree of doneness of meat. Time and temperature guides can be used to ascertain doneness, but cooking time is affected by fat, bone and moisture content and the shape and size of the cut. The basic types of meat cookery follow.

***Broiling:*** Broiling is recommended for all tender cuts and for best results:

- Set the oven for broiling
- Place thin cuts of meat on a rack at a distance from the heat equal to two times the thickness of the cut plus 2.5 cm
- Broil steaks, chops or patties for approximately one-half the desired cooking time before turning
- Season and serve at once.

***Pan-broiling:*** Pan-broiling is recommended for tender cuts suitable for broiling. For best results:

- Place meat in a hot frying-pan or on a griddle
- Do not add fat or water
- Cook slowly over moderate heat, turning occasionally
- Pour off or remove fat as it accumulates
- Brown meat on both sides
- Avoid overcooking.

***Roasting:*** Roasting is recommended for large, tender cuts. Some beef cuts suitable for roasting are rib and top sirloin roasts. For best results:

- Season with salt and pepper as desired
- Place the meat, fat side up, on a rack in an open shallow roasting-pan
- Insert a meat thermometre so that the bulb is in the centre of the largest muscle without touching bone.
- Add no water and do not cover
- Roast at oven temperature of 176°C to desired internal temperature.

Meats are usually cooked to degrees of doneness as follows:

| | |
|---|---|
| - Rare | 60°C |
| - Medium | 71°C |
| - Well done | 77°C |

***Pan-frying:*** Pan-frying is usually recommended for tender cuts 2.5 cm thick or less. For best results:

- Place meat in a hot frying-pan or on a griddle
- Fat may be added
- Cook slowly over moderate heat, turning occasionally
- Allow fat to accumulate
- Brown meat on both sides
- Avoid overcooking.

***Braising:*** This method is best used for less tender cuts such as beef round or chuck steak, pot roast, stew or short ribs. For best results:

- Use a heavy pan
- If desired, brown meat slowly on all sides with sufficient fat to keep meat from sticking

- Season with salt, pepper, herbs or spices
- Add a small amount of liquid
- Cover tightly
- Cook slowly over low heat on a stove burner or in a moderate oven until meat is tender.

Braising with large cuts is often called pot-roasting and with thin cuts may be known as Swissing.

***Simmering:*** This method consists of cooking a small amount of meat with a large amount of water. For best results the container should be tightly covered and the meat cooked slowly below the boiling point until tender. This method is used for the production of soups to which vegetables, grains or pasta products may be added.

## Meat as Raw Material, Non-meat Ingredients and Basic Techniques in Further Processing of Meat

### *Types of Animal Tissue Suitable For Meat Processing*

Meat is defined as those animal tissues which are suitable for use as food. These are the main soft tissues of the carcass: muscle, mainly skeletal (30– 65 percent), fatty (10–45 percent) and connective tissues. Other animal tissues used as food, and also to some extent in meat processing, are the internal organs including the blood.

### *Muscle Tissue*

The structural unit of muscle is a specialized cell, the muscle fibre, which constitutes 72–92 percent of the muscle volume. The membrane surrounding the muscle fibre is called the sarcolemma and the intracellular substance the sarcoplasm.

The muscle fibre is composed of many myofibrils, which consist of thick and thin filaments (myofilaments). The special arrangement of these and the bands of myofibrils give the fibre a striated appearance under a microscope (cross-striated muscle). The filaments consist almost entirely of the myofibular proteins actin (thin 20– 25 percent) and myosin (thick 50–55 percent).

Although they make up only 7 percent of muscle weight, they are mainly responsible for a very important property of meat, its ability to retain water and bind added water (water-holding capacity, WHC). The water-holding capacity is of particular importance in meat processing.

### *Connective Tissue*

Connective tissues are distributed throughout all body components -skeleton, skin, organs, fat, tendons and muscles. There are three kinds of connective tissue fibre: collagen, reticulin and elastin. Collagen constitutes 20–25 percent of total protein, and has a major (negative) influence on meat tenderness. Skin (from pigs only) has excellent swelling and binding abilities owing to its high collagen content. It is therefore ideal for meat products such as emulsion-type cooked sausages provided it is properly scalded, completely dehaired, usually singed, scraped, washed and de fatted.

### *Fatty Tissues*

The main fatty tissue deposits are in septa between muscle bundles (intramuscular fat), in spaces between muscles (intermuscular) and between skin and muscles (subcutaneous or backfat). Fat depots are also found around internal organs. The main depot is found around the kidneys (perirenal, leaf or kidney fat). Fatty tissues can be graded as "firm" (backfat, jowl and brisket) and "soft" fatty tissues (leaf perirenal fat) depending mainly on their connective tissue content.

### *Internal Organs*

Depending on local regulations and eating habits, the following are commonly used in sausage manufacture:

- Heart after removing the pericardium is used as any other kind of meat.
- Liver is used for making various types of liver sausage and paste, because its proteins have high emulsifying capacity.
- Tongue trimmed of all the hyoid bones, tonsils, and mucous membranes, can be cured and dried whole, used to make meat batter, or cured and canned (ox or pork tongues).
- Lungs. Beef lungs can be used to make cooked sausages. Pork lungs are frequently not fit for human consumption as they can be contaminated by scalding water.
- Kidneys are often contaminated to a certain extent with heavy metals or other residual substances and the consumption in higher quantities is not recommended in some countries.
- Tripe is the rumen and reticulum of ruminants, opened and rinsed. All the dark tissues (internal linings) must be removed by cooking (62– 65°C).

- Stomach of pigs, properly cleaned, is used as a natural casing for cooked sausages.
- Intestines are mainly used as casings for various sausages.
- Blood is highly perishable and must be handled carefully to avoid contamination during collection. To prevent coagulation blood is either defibrinated or a solution of sodium citrate 1.6 percent or phosphate 1 percent is added. Blood plasma obtained by centrifuging should be cooled as quickly as possible to 0°C. Whole blood is used to make blood sausage, liver sausage, and blood pudding. Blood plasma can be used for meat emulsions (batter).

### General Remarks

All raw materials must be fit for human consumption. After inspection, final dressing, removal of condemned and dirty parts and washing, all meat and organs must be immediately hung on hooks and moved to a cooler to await processing.

Carcasses may be fully or partially boned before chilling provided high hygienic standards are rigidly observed. A high degree of skill and special organization of labour is required. If small-scale producers cannot chill the carcasses, they may use hot-boned meat for sausage production or meat batter. Hot-boned meat has a high WHC so the use of phosphate is avoided. However, beef must be processed within four hours and pork within one hour of slaughter.

*PSE and DFD meat.* Pre-slaughter stress may result in abnormal undesirable muscle conditions called "pale, soft and exudative" (PSE), and "dark, firm and dry" (DFD). PSE meat is frequently found in pork caused by a sudden stress before slaughter. Glycogen levels are raised in response to the stress so that post-slaughter glycolysis is elevated leading to a build-up of lactic acid and a rapid fall in muscle pH to below 5.8 within one hour. This results in partial protein denaturation reducing WHC and increasing drip loss.

A prolonged period of stress prior to slaughter such as fighting during transport and lairage causes exhaustion and the depletion of glycogen reserves. Post-mortem glycolysis and lactic-acid production are therefore reduced, the pH falls slowly and protein degradation is reduced. The resulting DFD meat which is found in pork and beef has a high WHC but spoils very quickly because the high pH and dry surface favour bacterial growth.

### Simple Standards of Different Qualities of Meat Used As Raw Material For Different Product Qualities

Manufacturers must aim for uniform quality of their meat products. To attain this, raw materials must be standardized for different qualities. Goodquality carcasses are usually divided into primal cuts (ham, shoulder, loin, neck, etc.).

The remainder of the carcass and trimmings from the primal cuts are standardized into different qualities of meat. Poorer-quality carcasses are used entirely for processing after being deboned and trimmed.

Basic parametres for simple quality standards are size and shape of meat pieces, amount of visible fatty and connective tissues, and chemical composition. Meat must not contain skin, lymphatic glands, particles of bones, bristles, large blood vessels or blood clots.

### Different Qualities of Meat

*First quality* (Meat I). Meat pieces of relatively uniform size and shape, trimmed of connective tissue, with about 8 percent visible fatty tissue obtained from larger primal cuts, mostly hindquarter.

It is used to make meat batter for sausages and high-quality canned products.

*Second quality* (Meat II). Meat pieces of irregular size and shape, partially trimmed of connective tissue with about 15 percent visible fatty tissue, obtained mostly from forequarters. It is used to manufacture meat batter, or medium-quality meat products.

*Lean trimmings* are small irregular pieces of meat, with pervading connective and fatty tissue (about 25 percent), obtained during deboning and trimming of primal cuts, Meats I and II, and meat parts of the head and flank.

They are used to make meat batter for all kinds of medium-and lowquality cooked sausages.

*Fatty trimmings* are meat pieces containing about 50 percent visible fatty tissue, derived from all trimming operations. They are used as the fatty ingredient of meat batter of medium and low quality.

### Standards of Fatty Tissues

***Ruminants:*** *Firm (external) fatty tissues* are trimmed from any part of the carcass and hump of zebu and are used as the fatty component of the meat batter or sausage mixture (mutton, goat, etc.) or all-beef higher-quality sausages.

*Soft (internal) fatty tissues* are perirenal and sacral in origin and are not generally used in sausages.

### Pork

*Jowl* (firm) contains much muscular (even up to 30 percent) and connective tissues. It is suitable for manufacturing dry sausages, summer sausages, salamis and emulsion-type sausages of the highest quality.

*Back fat* (firm) is used for semi-dry and dry sausages, and in frankfurter sausages of the highest quality.

*Side fat* (firm) contains about 60 percent visible fatty tissue. It is used to manufacture medium-quality cooked sausages made of uncooked or precooked materials.

*Belly fat and leaf* (soft) are used to make cooked sausages of low quality.

### Chemical Parametres

Typical quality standards based on average chemical composition are shown in table.

## Salt, Curing Agents, Common Spices, Additives and Natural Smoke

### Common Salt

Common salt (sodium chloride, NaCl, salt) which may be extracted from sea water (sea salt) or mined (rock salt), has three major effects upon meat.

*Flavour enhancement* of meat and meat products. The salty taste of a meat product depends on the relative amounts of salt and water. Typical ranges of salt concentration for various products are shown. Products with less water require higher levels of salt concentration to achieve the same degree of saltiness.

*Functional properties of meat proteins*. Depending upon its concentration salt can increase or decrease the WHC of a meat product. The dehydrating effect of salt is used for meat drying (lowering WHC).

The opposite effect of increasing WHC is very important and results from the swelling and solubilizing of the muscle proteins (actin and myosin).

***Table:*** *Approximate chemical composition of different qualities of raw materials*

| *Raw materials* | *Water* | *Fat* | *Protein* | | |
|---|---|---|---|---|---|
| | | | *Muscle* | *Connective tissue* | *Total* |
| Meat I | 71 | 10 | 16 | 3 | 19 |
| Meat II | 63 | 20 | 12 | 5 | 17 |
| Lean trimmings | 53 | 33 | 10 | 2 | 12 |
| Fat trimmings | 30 | 60 | 7 | 2 | 9 |
| External beef fat | 27 | 67 | 1 | 5 | 6 |
| Internal beef fat | 5 | 93 | - | 2 | 2 |
| Jowl | 17 | 78 | 3 | 3 | 6 |
| Back fat | 8 | 90 | - | 2 | 2 |
| Side fat | 32 | 60 | 7 | 1 | 8 |
| Soft fat | 5 | 93 | - | 3 | 3 |
| Pork skin | 55 | 15 | 0 | 30 | 30 |

***Table:*** *Typical concentration of salt in some meat products*

| *Finished Meat Products* | *Range of Salt Concentration (%)* |
|---|---|
| Sausages | |
| fresh and cooked | 1.6–2.2 |
| dry, small-diametre | 2.4–3.8 |
| dry, large-diametre | 3.0–5.0 |
| Hams | |
| cooked | 2.6–3.2 |
| dry | 2.5–6.0 |

*Preservation.* Salt is one of the most important food additives in food preservation. The salt concentration determines what types of microorganism, if any, can grow by dehydrating or by lowering the amount of water available for growth.

## Curing Agents

*Nitrates.* Sodium ($NaNO_3$) or potassium nitrate (saltpeter, $KNO_3$) allow cured meat colour to develop in products where drying is a long-term process. Nowadays, they are used less frequently because to be effective they have to be reduced to nitrites under the influence of bacterial enzymes, and this is a time-consuming process.

*Nitrites* are indispensable for meat curing, and no substitute has yet been found. Sodium nitrite ($NaNO_2$), a toxic substance, can be fatal even in small doses. For this reason they are often mixed with common salt at a concentration of about 0.6 percent (so-called “nitrite salt”) when used for curing. If excessive levels of nitrite are accidently reached the accompanying salty taste will be rejected by the consumer, thereby preventing nitrite poisoning.

The maximum amount of nitrite permitted in finished meat products is usually 200 ppm (parts per million, or mg per kg), or may be less subject to the type of meat product or country legislation.

Saltpeter can be added to the nitrite salt at a concentration of 1 percent and used for curing dry hams and dry sausages.

***Table:** Typical amounts of nitrite and nitrate in cured products*

| Curing agents | Amount of nitrite or nitrate in cured-meat products |
|---|---|
| Nitrite salt (99.4% | all-meat products, 100 ppm as nitrite dry |
| NaCl + 0.6% $NaNO_2$) | hams, 150 ppm as nitrite |
| Saltpeter ($KNO_3$) | dry sausages, 100 ppm as nitrate low-sodium products, 100 ppm as nitrate |
| Nitrite salt + saltpeter | dry hams, 600 ppm as nitrate |

Three processes in meat curing are due to the effect of nitrites:

*Cured-meat colour development* is achieved when the muscle pigment (myoglobin) in an acid environment combines with nitric oxide (NO) (formed from nitrite) to form NO myoglobin.

This reaction is affected by temperature, pH and oxygen-reducing agents. NO-myoglobin is relatively resistant to light and oxygen and, most importantly, it is heat stable.

Thus, cured cooked meat and meat products maintain a bright red colour in contrast to uncured meat which turns grey after cooking.

Nowadays it is considered that 3–50 ppm is sufficient to achieve colour in cooked sausages.

*Cured-meat flavour development* is based on various reactions between nitrite and the meat component. Typical flavour of cured-meat products is achieved with 20–40 ppm nitrite.

*Preservative effect*. Even in small doses (80–150 ppm), nitrite prevents the growth of numerous micro-organisms, and food-poisoning bacteria (*Clostridium botulinum*, salmonella, staphylococci, etc.).

However, the effect of nitrite on shelf-life or prevention of food-poisoning bacterial growth must not be overestimated and decreases with increasing storage temperature.

## Common Spices

*Spices* act on the salivary and gastric glands to promote secretion, stimulating appetite and improving digestibility of meat products. Their use varies from country to country depending on the climate, customs and eating habits. There are spices whose taste and smell remain unchanged even after exposure to high temperatures (chilies and sage). Less resistant are cardamom, clove, pepper, rosemary and thyme, and the least heatresistant are coriander, mace, marjoram, nutmeg, allspice and ginger.

## Useful Additives

*Phosphates* are used to restore WHC to chilled meat, approximately to the same level as hot-boned meat. Certain countries forbid phosphates, whereas some allow their use only where there is a proven technological effect. Where permitted they should be restricted to 0.3–0.5 percent of the sausage mixture weight. Phosphates break down actomyosin into actin and myosin, which can be solubilized by salt to increase the WHC. This effect is retained even in cooked products, increasing the yield.

*Ascorbic acid* (vitamin C) and its salts (sodium ascorbate) contribute to the development of cured-meat colour.

Sodium ascorbate is used in the manufacture of cooked sausages, made from uncooked or precooked raw materials. Ascorbic acid used is at a concentration of 0.03–0.05 percent, whereas sodium ascorbate is added at a concentration of 0.07 percent. Ascorbic acid is a strong reducing agent, enabling quicker formation of the NO-myoglobin so that less nitrite is needed, and it inhibits the formation of an undesirable colour in cured-meat products. It must not be added to, or mixed with nitrites, because they will be broken down instantly and will become useless for curing. Thus, the nitrite salt must be added to meat at the very beginning, whereas ascorbic acid is always added at the end of comminution.

Ascorbic acid decomposes rapidly especially in a humid warm environment. Its salt (sodium ascorbate), being more stable, is often used in sausage production, as is erythorbic acid and its salt (sodium erythorbate).

*Glutamates.* Monosodium glutamate and other salts of glutamic acid are substances which improve the flavour of meat products, and are usually added in concentrations up to 0.2 percent.

***Table:** List of common spices used in sausage seasoning (g/kg sausage mixture)*

| ***Spice*** | ***Drysausages*** | ***Made of raw Material*** | |
|---|---|---|---|
| | | ***Precooked*** | ***Uncooked*** |
| Allspice | 0.5–1.0 | 0.3–0.5 | 0.1–1.0 |
| Anise | - | 0.1–0.3 | - |
| Bay leaves | 0.01 | 0.05–0.1 | - |
| Caraway seed | 0.2–1.5 | 0.2–0.7 | - |
| Cardamon | 0.1–0.5 | 0.2–0.5 | 0.3–0.5 |
| Cinnamon | - | 0.05–0.2 | - |
| Cloves | 0.2–0.5 | 0.1–0.3 | 0.3–0.5 |
| Coriander | 0.3–0.5 | 0.3–1.0 | 0.2–1.0 |
| Cumin seed | 0.1–0.5 | 0.2–1.5 | 0.2–0.7 |
| Ginger | 0.1–0.3 | 0.3–1.0 | 0.3–0.5 |
| Mace | 0.1–1.1 | 0.3–1.0 | 0.3–2.0 |
| Marjoram | 0.2–0.3 | 0.5–2.0 | 0.2–1.0 |
| Nutmeg | 0.2–0.5 | 0.3–1.0 | 0.3–1.0 |
| Paprika | 0.1–1.0 | 0.4–1.0 | 0.5–2.0 |
| Pepper | 1.0–4.0 | 1.5–3.0 | 1.0–4.0 |
| Sage | 0.05–0.5 | 0.1–0.3 | 0.05–0.5 |
| Tarragon | - | 0.1–0.3 | 0.05–0.5 |
| Thyme | - | 0.1–0.3 | 0.1–0.3 |
| Garlic | according to taste | according to taste | |
| Onion | - | according to taste | |
| Mustard seed | 0.5–2.0 | 0.1–0.2 | |

## Natural Smoke

Natural smoke is a very complex mixture, consisting of a great number of compounds, and is obtained by controlled combustion of moist sawdust at low temperature. Sawdust from hardwoods is most commonly used to generate the smoke. Nowadays, it is considered that optimal smoke composition is obtained at temperatures of 300–500°C.

Smoke consists of gases (phenols, organic acids, carboniles and other compounds) and particles (pitch, tar, ash and soot). Gaseous components penetrate into a product through the casing to a certain level, and react with other components of meat products. Other components are deposited on to the surface. Smoke provides typical flavour and distinctive colour, and hardens the surface of the meat product.

### General Remarks

All substances which are added to meat products must have food grade purity. They should not contain any food-poisoning bacteria, so must be treated according to the highest hygienic standards. It is important to keep them in properly closed containers or intact packages, away from any dampness and dust. They are usually kept in special, dry premises away from the workshop, in which they can be pre-weighed, blended and packed into plastic bags in the proportions required for sausage formulations. The nitrate must be kept under lock and key.

Dosage by hand of any non-meat ingredient is not allowed. The only correct way is with scales which must be checked occasionally for accuracy. One of the most serious consequences of failure to protect all non-meat substances is contamination with dirt, excreta from rodents, birds or other animals and infestation with insects.

### Principles of Heat Treatment For Non-sterilized Products

During processing many meat products are subject to specific heat treatment. The first task of heat treatment is to reach satisfactory shelf-life by reduction of micro-organisms. The second task is to obtain desirable organoleptic characteristics, to preserve nutritive value and improve digestibility of the product.

### Reduction of Micro-organisms

Bacteria are destroyed if exposed to sufficiently high temperatures for long enough. There is a direct relationship between bacteria survival and time of exposure to temperatures. As an example, if 10 000 000 bacteria (per ml) suspended in broth are exposed to heat (70°C), after the first five minutes 1 000 000 will survive (90 percent are destroyed), after the next five minutes the number of surviving will be 100 000 (again 90 percent are destroyed), and so forth.

This tenfold reduction in bacterial numbers between fixed time intervals is called decimal reduction. The time interval for decimal reduction varies between different bacteria and depends on the temperature applied. The number of bacteria present in a meat product just before the heat treatment (initial number) should be as low as possible so that a shorter time or lower temperature is needed to achieve a satisfactory shelf-life for the product.

As sausage fillings as well as most other meat products represent a very good medium for bacterial growth, they should immediately be exposed to heat treatment in order to prevent bacterial growth. It is also important to perform all operations as quickly as possible, and to maintain the highest hygienic standards so that the initial bacterial count remains as low as possible. The manufacturer must always bear in mind that bacteria grow very fast. Their number may be doubled every 20 minutes.

### Organoleptic Changes

These are caused by heat treatment (doneness, flavour, firmness, consistency and cured-meat colour development) are time-temperaturedependent processes. The basic effect of the heat treatment is coagulation of meat proteins. Between 70° and 80°C the majority of meat proteins are completely coagulated, forming a structural matrix which entraps fat and water droplets released during heat treatment. With coagulation of meat proteins, WHC is decreased and the meat loses a certain amount of water (thermal weight loss). Those structural changes of proteins are responsible for characteristic firmness of heat-treated products. Frankfurters have an elastic firmness, and on reheating prior to consumption become even more firm. Products containing connective tissue become more tender owing to solubilization of the collagen (gelling). Products such as paste which are in a liquid state prior to heating change for more viscous and spreadable consistency. Sausages made from larger meat pieces also attain a characteristic consistency. The exudate released during massaging or tumbling coagulates and binds the pieces of meat.

***Table:*** *Influence of temperature and time of heat treatment on cured-meat colour development in bologna (diametre 90 mm)*

| ***Temperature of Heating Medium*** | ***Time Necessary to Develop*** | |
|---|---|---|
| | ***Pinkish(not stable)*** | ***Red colour(stable)*** |
| (°C) | (min) | |
| 70 | 90 | 110 |
| 75 | 75 | 90 |
| 80 | 60 | 75 |
| 100 | 30 | 45 |
| 110 | 15 | 30 |

Such products (cooked hams, ham sausages) become sliceable. Heat treatment also makes products chewable.

In addition to these physical changes resulting from heat treatment, some biochemical reactions also take place which contribute to the typical flavour of heat-treated products. Many compounds present in the raw material are broken down by heat to produce the characteristic flavour. Fatty components give the meat product a distinctive flavour, specific to the animal species from which fat is obtained.

Cured-meat colour is not stable unless the raw product is heat-treated.

All these changes occurring during heat treatment give the product its typical overall sensation of doneness.

### Duration of Heat Treatment

The duration of the heat treatment primarily depends upon the size and shape of the meat product. Treatment continues until the coldest point (innermost part) reaches the defined temperature. It is essential to check temperatures of the heating medium and of the innermost part of the products and to observe exact times.

***Table:*** *Heat treatment of different meat products*

| ***Product*** | ***Temperature*** | |
|---|---|---|
| | ***Heating medium*** | ***In product*** |
| | *(°C)* | |
| Fresh sausages, hamburgers | 150–350 | 65–110 |
| Sausages made of uncooked raw material | 75–85 | 70–75 |
| Sausages made of precooked raw material | 80–98 | 85–90 |
| Canned cured hams, shoulders and loins | 75–85 | 70–75 |

### Types of Heat Treatment

*Heat treatment by dry heat* is performed in special ovens (roasting) or on a gridiron (grilling). Meat loaves and similar products are roasted. Meat patties, hamburgers and fresh sausages are grilled.

*Heat treatment by hotwater* is the most common and is usually performed in large cooking vats. When applied to canned hams or ham sausages, it is called pasteurization.

*Heat treatment by steam* is one in special steam cabinets and is in particular applied in cases when treatment in cooking vats is not desirable because of substantial losses of aroma and flavour of the products into the cooking water.

### Frequent Faults Committed During Heat Treatment

*Undercooking* may be due to the temperature being too low, the time of treatment being too short, or both. The consequences are in a reduced shelflife due to spoilage and the possibility of food-poisoning. Spoilage results in a bad smell, discolouration, softened consistency and souring.

*Overcooking* results from either too high a temperature or too long a duration or both. The consequences are increased cooking losses, fat separation in some sausages, undesirable changes of flavour and a softer consistency.

### General Remarks

In order to keep the initial bacterial numbers as low as possible it is important to keep the temperature of the raw materials as low as possible before heat treatment. After filling sausages and mixing other meat products heat treatment must be applied immediately.

## Raw or Cooked Meat Products Fabricated From Entire Meat Pieces

### Raw and Dry Meat Pieces

These are prepared from primal cuts with or without bones. They are cured and dried without heat treatment and must be produced from wellrefrigerated carcasses. This process is based on two principles: the stabilization of the product by decreasing water activity and adequate ripening. Decreasing water activity (= amount of water available for microbiological growth) is achieved by penetration of salt into the meat pieces and consequent dehydration. With 3–5 percent salt penetration, 5–9 percent of the water will migrate out of the meat. Dehydration continues during drying and ripening and prevents bacterial growth. Nitrite, which is either added directly or formed from nitrate, also contributes to this bacteriological effect. If these preservative conditions are completed, the finished dry product will be bacteriologically stable and need not be cooked before consumption. At the same time, these processes destroy possible live trichinae in pork. Repeated rearrangement during curing of stocked meat cuts is important to eliminate pressed-out water.

Smoking, drying and ripening contribute to the development of very pleasant flavour, which is due to biochemical changes taking place in muscle and fatty tissue. First lipolysis takes place in which complex molecules of fat are broken down under the influence of enzymes, not only those normally present in fatty tissue, but also those of bacterial origin (especially Micrococceae). Second, oxidation is initiated by higher temperatures. Intensive oxidation of fat can cause undesirable rancidity.

All these reactions taking place during processing develop very slowly, because the useful enzymes are enclosed in the cells. Drying and ripening must therefore be long-term processes for the enzymes to be effective.

### Raw Dry Hams

Dry hams are uncooked, cured, dried and usually smoked pork legs.

*Raw material.* The legs are cut from the pork sides and quickly chilled to an internal temperature of 2°C within 18 hours. Legs having blood clots, PSE or DFD muscles are not suitable. The minimum weight of a deboned leg is 5 kg.

*Additives.* A mixture of salt (10 kg), nitrate (400 g), sugar (500 g) and spices, usually black and white pepper, sage and ginger, is used for dry curing. Instead of salt and nitrate, the so-called nitrite salt can also be used. In some traditional methods only common salt is used.

*Manufacturing.* In the cold season and at high altitudes hams may be dried under natural conditions, according to the traditional method. Because of air purity and better circulation, nitrate is not usually used, so a larger quantity of coarse salt must be added, giving the finished product a more salty taste.

After overnight chilling, legs are cut off between the aitch bone and femur. Feet are removed but the skin is usually left on, but may be removed from the upper third of the leg to improve salt penetration. The legs are rubbed with the curing mix, arranged in a wooden cask placed in a cool place (about 0°C) and hand sprinkled with the same mix. Salting lasts about three weeks. After a week they are salted separately and rearranged in another properly washed cask. After salting the legs are rinsed and the remaining salt is removed. They are tied and hung in a cool place for further drying and ripening for at least five to six months. The windows and doors should occasionally be opened and closed and ventilation provided. Legs should be rearranged so that they are all uniformly dried.

Smoking can be applied in the processing of dry hams. Its purpose is to give the product a typical flavour, distinctive colour and to harden it.

Smoking quality depends on many factors, but the most important are the smoke temperature at the product surface and air humidity. It is most important that the surface is dry enough before smoking starts. If smoking is a long-term process, drying occurs simultaneously but with a shorter process it must be done first in a drying room. In raw dry ham production only cold-smoking is used (below 25°C). Smoke density and duration depends on local preferences.

The industrial method uses special air-conditioned coolers, in which standardized conditions (temperature, relative humidity (RH), air ventilation and circulation) can be maintained all year round.

*Frequent faults committed during production.* Faults may arise due to the use of poor-quality raw material, inadequate manufacturing, unclean containers, and/or non-hygienic conditions. The most common are:

- Crust on the upperface (3–5 mm thick), of hard consistency and darker colour which arises during smoking caused by low air humidity (below 65 percent) and/or very strong air circulation.
- Cracks around the head of the leg bone of varying size and depth, due to careless deboning and excessive drying.
- Insufficient drying of certain parts can appear in large hams, seen as lighter areas with a softer consistency. It is caused by improper processing (especially during drying and ripening), crust formation (which prevents uniform drying of the inner parts), and/or insufficient air ventilation (especially when dealing with legs weighing more than 9 kg).
- Mould (yellowish or greenish) is caused by drying at high RH, due to poor ventilation and air circulation and/or mould contamination of equipment and workshop.
- Oversalty taste is the consequence of improper desalting.
- Rancidity arises from the oxidation of unsaturated fatty acids and in severe cases results in a yellow discolouration. High temperatures applied during smoking, drying and ripening, or poor-quality feed may be the cause.
- Spoilage within the product usually occurs around large blood vessels and is usually due to the raw material being

contaminated during deboning, trimming and/or curing in non-hygienic containers. The characteristic brownish colour and unpleasant smell are caused by anaerobes.

- Acarid findings show that products have not been properly protected against insects.

*Storage and shelf-life*. If the process is carefully followed, the dry-cured ham is very tasty, nutritive and bacteriologically stable. Stored at room temperature, but not more than 30°C, and RH between 55–65 percent, the shelf-life should be four to six months. Raw dry hams are ready for sale immediately after production and final control. They are sold as they are, or may be packed (half or small pieces) into plastic bags. All deboned meat pieces may also be sliced and packed under vacuum, in different consumer sizes, as trade demands.

The most common use of dry hams and all other uncooked meat products is in the form of very thin slices, for use in sandwiches or as starters. Bone-in products are used for preparing various cooked dishes.

### Other Uncooked Processed Meat Pieces

In addition to dry hams, many other uncooked dry processed meat pieces may be manufactured from different primal cuts of pork carcasses and other species. All parts of a pork carcass may be used (ham, shoulder, loin, neck, back fat, jowl, head, tongue, ears, spare-rib, shank, feet, tail, bones etc.). A long shelf-life is achieved by a combination of long drycuring (more than two weeks) and long cold-smoking, with simultaneous drying and ripening.

Dry-curing is always followed by repeated salt rubbing during weekly rearranging of meat pieces. Duration of salting or curing depends on size and on the coverage of skin or fatty and connective tissues.

All uncooked processed meat pieces must have: a regular shape without unnecessary cuts and rough spots; dry, clean surfaces; moderate cured salty taste and agreeable bouquet; and a water content in finished product not exceeding 30 percent. If the product has skin, it should be clean, light to dark gold-brownish colour, without bristles.

"Country pork shoulder" is processed in a similar way to raw dry hams, though with a shorter production time. The combined period for curing and salt equalization should be at least 25 days, and the total production time at least 50 days.

Dry neck can be made with or without the bones. If bone-in it is cut along the neck bones and first three vertebrae and the upper parts of the first three ribs (3 cm in length). Both types are dry-cured for a minimum of three weeks. Well-cured deboned necks can be filled firmly into fibrous casing or elastic nets. The thicker part must be tied and hung on a stick, and left dripping for five to six hours. When the surfaces are dry enough, coldsmoking can start and should last three to four weeks.

Dry loin can also be made with or without the bones. In both cases it must be cut between the third back bone and last lumbar vertebrae, with the corresponding ribs to a width of 3 cm. The external layer of fatty tissue must not exceed 0.5 mm. If all muscles except the long back muscle are trimmed off it is called "dry real loin".

Dry spare-ribs are the cured, smoked and dried lower half of the ribs, with muscles between them.

Dry head is the cured, smoked and dried half head of all kinds of small animal, totally deboned, without salivary and lymphatic glands and tonsils.

Dry shanks and feet are well-cleaned, singed, cured, smoked and dried sheep, goat, deer and pig feet.

Dry tail, usually cut off together with the aitch bone, is a cured, smoked and dried product.

Sheep, goat or small venison pastrami are cured, smoked and dried carcasses or sides, without the head and neck, either bone-in or partially deboned. In the first case, for better salt penetration, the largest muscles must be incised, the joints opened in the interior part, and the long bones broken longitudinally. In partially deboned products, all bones except the vertabrae are removed. In both cases, dry-curing agents are rubbed in and the carcasses or sides are spread in wide vats, to prevent creasing. Curing lasts from three to four weeks, depending on the thickness. During that time the product is recured twice, with repeated rubbing. After curing it must be flushed, tied with many loops and hung on a stick for stretching by placing two rods longitudinally in the shoulder and ham. Smoking lasts for at least a month to give a desirable gold-brownish colour and an attractive flavour.

Beef pastrami is dry-cured and smoked beef navel part or well-trimmed brisket.

The raw material must be rubbed on both sides with a mix of nitrite salt and saltpeter and left in curing vats for two weeks. After seven days, the product must be repacked and, if necessary, some parts rubbed again. After curing residual salt is eliminated by washing and the product is hung for dripping. Smoking starts when the surfaces are dry enough, at 54°C, with light smoke, for three hours.

The amount of smoke and the temperature are gradually raised until the inside temperature of 74°C is reached. The product is then removed from the smokehouse and hung at room temperature for a minimum of two hours before being placed in the cooler prior to sale.

### Cooked Hams and Cured and Cooked Meat Pieces

The majority of this type of processed meat is fabricated from pork, but there are also certain cured and cooked beef products.

Cooked hams are made either from entire hams (pork leg muscles, with or without shank) or reconstituted from leg muscles or parts of them. As a rule, hams are produced as boneless, skinless, shankless and defatted products.

Most often, they are packed in a can or plastic pouch sealed under vacuum. In both cases, pasteurization (cooking at temperature lower than 100°C) is done in the final container in which they remain until slicing. Being perishable they must be kept under refrigeration. The exception is with small packages (lighter than 1.4 kg) of hams which can be produced as commercially sterile products.

Beef hams are made from meat from younger cattle and the same process is applied as for the manufacture of cooked hams.

*Additives.* Water, salt, sugar, phosphate, nitrite and sometimes salts of ascorbic acid and soy isolates are the basic ingredients of curing brine. The amount injected depends on product quality. Water used for preparing curing brine must meet the standards of potable water. All ingredients should be well chilled.

Data presented in Tables 10 and 11 show that injecting 12, 31 or 36 percent curing brine of appropriate composition will give the desirable amount of salt (2.6 percent) and nitrites (0.016 percent) in the finished product. The amount of other additives may vary with local regulations.

*Manufacturing.* The highest standards of hygiene must be maintained at all stages. The processing steps are:

Brine injection. For small production one-needle brine injectors are sufficient. The modern method uses pickle injectors with a large number of needles, providing a more even distribution of the curing brine. Pressure during injection can be regulated, as well as the speed of the conveyor providing the material. These multi-needle injectors allow more brine to be injected into products than the traditional method.

The pressure of injection varies from 1 to 2 atm, depending on the number of needles used and the amount of brine to be injected. For high brine injection rates (20 percent or more) it is better to inject twice under low pressure than once under high pressure, to avoid the formation of "lakes" of brine in the muscles. The weight of meat must be checked before and after injecting. If the percentage of injected brine is less than that required, it can be made up by adding it in the massaging vat.

***Table:*** *Ingredients necessary for preparing curing brine for manufacturing regular hams (I), soy- and water-added hams (II) and ham imitations (III)*

| ***Ingredients*** | ***Quality of hams*** | | |
|---|---|---|---|
| | I | II | III |
| | | *(%)* | |
| Tap water | 70.43 | 82.11 | 84.59 |
| Nitrite salt | 21.73 | 8.49 | 7.26 |
| Dextrose | 4.16 | 1.61 | 1.39 |
| Phosphate | 3.60 | 1.39 | 1.19 |
| Soy isolate | - | 6.40 | 5.55 |

Mechanical treatment of meat after brine injection has become a recommended process for larger-scale production, not only for cured entire meat pieces but also in the production of reconstituted smaller pieces. During mechanical treatment, muscular cells are ruptured so that myofibrilar proteins quickly come in contact with the brine and swell. Exudate which consists of cell juices, brine, meat particles and salt-soluble proteins acts like a "glue" to bind meat pieces together. Processed ham can therefore be easily sliced without falling apart.

The advantages of mechanical treatment are shorter curing time, better yield, less cooking loss, improved tenderness and better sliceability. Duration of the mechanical treatment depends on the machine and usually lasts from nine to 24 hours at 2–4° C.

There are two kinds of machine used for mechanical treatment:

Tumblers are containers in the form of a cylinder rotating around the axis.

The speed of rotation is 10–20 revolutions per minute (rpm) and the total number of revolutions should be 4 000–8 000, depending on the size of meat pieces and machine size. As a rule, tumbling lasts for about 18 hours, with five-to ten-minute tumbling intervals followed by 20- to 25-minute pauses.

***Table:*** *Raw material composition of regular hams (I), soy-and water-added hams (II) and ham imitations (III)*

| ***Ingredients*** | | ***Quality of hams*** | |
|---|---|---|---|
| | III | I | II |
| | | (%) | |
| Whole muscles | 85.00 | 66.00 | 0.00 |
| Chopped muscles | 3.00 | 3.00 | 64.00 |
| Total meat | 88.00 | 69.00 | 64.00 |
| Water | 8.45 | 25.45 | 30.45 |
| Nitrite salt | 2.62 | 2.62 | 2.62 |
| Dextrose | 0.50 | 0.50 | 0.50 |
| Phosphate | 0.43 | 0.43 | 0.43 |
| Soy isolate | - | 2.00 | 2.00 |
| Total brine injected | 12.00 | 31.00 | 36.00 |

Tumbling under vacuum is recommended to avoid foaming and to improve colour stability.

Massaging vats have handles placed horizontally on the vertical axis propelled by the electrical engine which squeeze the larger pieces of meat. The duration of massaging, time intervals and total time are similar to tumbling. The use of a vacuum is not possible with these machines.

Heat treatment (pasteurization) of the product in plastic pouches, cans or ham boilers, is done either in an autoclave or a cooking vat in water at 80-85°C. Cooking lasts until the temperature at the geometrical centre (the coldest point of the product) reaches 70–75°C. Containers are cooled under running water until the temperature in the centre of the product falls to 35°C, when they are moved into coolers at 2–4° C.

*Frequent faults committed during production*

- Improper trimming of connective tissue which contracts during thermal treatment resulting in an uneven surface and holes in the cross-section.
- Bad sliceability of hams (pieces fall apart) is the result of insufficient mechanical treatment and/or insufficient exudate.
- Holes in the cross-section result from insufficient vacuum used during can stuffing.
- Uneven colour of slices due to incomplete trimming of PSE muscles whose pale colour persists to the finished product.
- Excessive cooked-out juice results from incorrect brine composition (insufficient salt or phosphate) or improper mechanical or heat treatment.
- Overcooking beyond the temperature required for pasteurization increases the amount of cooked-out juices, and gives the product surface a softer consistency.
- Undercooking due to insufficient heat treatment is the greatest and most serious fault. The most important consequences are shorter shelf-life and/or rapid spoilage. Less important are harder consistency and uneven colour of slices.

*Storage and shelf-life*. Canned cooked hams should be kept at 5°C to give a shelf-life of six months.

Thin slices of ham are used for sandwiches or as starters. Smaller slices are used for ham and eggs.

Other products such as canned cured shoulders, loins, necks, etc. are manufactured in a similar way.

## Meat Products Fabricated From Comminuted Meat, Fat and Offal

### *Comminuted Meat Products*

Comminution is the mechanical process of reducing raw materials to small particles. The degree of comminution differs among various processed products and is often a unique characteristic of a particular product ranging from very coarsely comminuted, to finely comminuted, to form an emulsion.

*Machines for comminuting*. The range and quality of finished products prepared from comminuted meat depend on the skill of personnel and the equipment available. A minimum layout should include grinder, cutter, emulsion mill and ice-maker. Machines must be designed for easy cleaning.

All surfaces in contact with products must be smooth, free from pits, crevices and scales. Machines must be constructed either of stainless steel, or heat-resistent, non-toxic plastic material. All machines and tools must be carefully cleaned several times during the working day and disinfected at the end of the day. Manufacturer's instructions about the use and the maintenance of the machines must be strictly followed.

The grinder is usually the first machine used in the comminution of sausages. For non-emulsion-type sausages, grinding is often the only form of comminution. It is used to cut the raw material and thoroughly mix the ingredients. Meat is pushed along a worm-screw and then through perforated plates. The holes of the grinder plates vary both in size (2–30 mm) and shape. If the plates and knives are not kept in good condition, and particularly if they are not sharp, meat will be overheated, become greasy and lose its binding ability.

The cutter is the most important comminuting machine, for simultaneous comminution and mixing. Meat revolves in a bowl and passes through a set of knives mounted on a high-speed rotating arbor in a fixed position. The meat is guided towards the knives by a plough fixed inside the bowl. There are usually two speeds each for the bowl and the knives. The knives can differ in size and shape from rectangular to round. There can also be a special device for charging and discharging the bowl.

To eliminate heating of the batter caused by friction, ice water is added. This is better than ice flakes alone. To avoid excessive heating, properly sharpened knives must be used and the clearance between the knives and the bowl should not exceed 0.7 mm. A thermometre is mounted on the cover of the bowl to monitor the temperature of the meat batter. Modern cutters can operate under vacuum, which improves the colour and other properties of the finished meat products.

Depending on the meat-particle size desired, it is possible to produce a satisfactory comminuted meat product using only the cutter. For very fine products, such as frankfurter or bologna, it is often preferable to pass the emulsion obtained in the cutter through an emulsion mill.

The emulsion mill is operated by one or more rotating knives, revolving at extremely high revolutions, pulling the sausage mixture through one or more plates. Emulsion temperature rises by up to 3°C on each pass through the emulsion mill.

The ice-maker freezes water on the spot into ice flakes.

*Tools.* There are many tools used in meat comminution.

Trays, metallic or plastic, are used to keep, transport and weigh small quantities of the various raw materials and for washing dismantled parts of machines.

Vats, metallic or plastic, are used to hold larger quantities of raw materials (up to 200 1).

Container trolleys are movable containers used to keep and transport raw materials from the cooler to the workshop, or between machines.

Filling tables are metallic tables upon which the filled sausages run from the filler. They are made with raised rims, except under the filler.

Sticks are wooden or preferably metallic equipment for hanging up linked sausages (individually or in links), or meat pieces, for dripping, smoking, heat processing and storage.

Cages are metallic structures in which loaded sticks are transported for further processing (smoking and heat processing).

Balance and scales are used for weighing various raw materials, salt, curing agents and all additives. They are also used for checking product weight before and after each stage of processing to control weight loss.

Thermometres are used to check product temperature during comminution.

## Basic Technology of Meat Comminution

Comminution is a procedure which modifies the properties of fresh meat, so that the finished product consists of small meat and fatty pieces, or it can be finely subdivided to obtain a meat batter. This meat batter is a viscous mass, with many properties of an emulsion. A major problem in manufacturing emulsion-type sausages is the tendency of fat to be separated during heat treatment. The sausage batter, at least in part, is formed of a “fat/water emulsion” (meat emulsion or, better to say, dispersion), employing salt-soluble proteins of meat as emulsifying agents.

Comminution, the basic method for sausage making can be: coarse, to produce non-emulsified sausages like salamis and summer sausages, or fine, to produce emulsified sausages like frankfurters, bologna, etc.

*Coarse comminution* is used to manufacture sausages with a coarse texture with small pieces of meat and fatty tissues visible. For Meat I and for Meat II a 15- to 25-mm grinder plate is used while trimmings and fatty tissues are passed through a 2- to 6-mm plate so that the connective tissues are less visible and more digestible. Meat I and Meat II are passed through the grinder first, followed by the trimmings and finally the fatty tissues. This avoids fat separated from the fatty tissues greasing the surface of meat pieces thereby decreasing their binding ability. The meat should be well refrigerated (2–3°C) and firm, so that it will be cut cleanly by the grinder knives and not squeezed.

Correctly pre-ground meat is transferred into the mixer or mixed manually for formulation. Nitrite salt and phosphate are added and the mixing starts. All other non-meat components are added during mixing, finishing with ascorbic acid. When the mixture is uniform the sausage mixture is formed.

*Fine comminution* used to manufacture emulsion-type sausages is performed in a high-speed cutter. The knives must be kept thoroughly sharpened and properly adjusted to the bowl. During comminution bundles of fibres, myofibrils and filaments are separated and disrupted and the size of meat and fatty tissues is continuously reduced. The muscle tissue, having a stronger structure than fatty tissue, determines the total comminution time required.

Heat produced during comminution helps the salt-soluble proteins to concentrate at the fatglobule surface and come in contact with water from the protein matrix. Higher speeds produce a better distribution of fat due to the temperature rise but excessive temperatures (more than 15°C in pork and 18°C in beef) can partially destroy the protein membrane which surrounds fat particles, and cause a breakdown of the sausage mixture. Unprotected melted fat migrates to the surface, forming easily visible fat pockets between the sausage surface and the casings.

The melting point of fat is the parametre for determining the maximum temperature for the meat batter. Overcomminution can also provoke breakdown. As the fat particle size continuously decreases, there is a proportional increase in the total surface of fat particles requiring proteins. Generally, as particle size decreases more meat and meat-extender proteins are required to form a heat-stable emulsion.

*Frozen and chilled meat for different products*. Frozen meat must be sliced in a frozen-meat slicer prior to grinding. The block of meat is placed upon the carriage and passed under a large knife which cuts the meat into large slices or smaller meat flakes.

Frozen meat is not suitable for coarse comminuted sausages but may be used in manufacturing fine comminuted meat products. So that the optimal temperature (3–5°C) for extraction of salt-soluble proteins is reached, only water without ice is added in the cutter.

Some meat extender must be added to bind juices from the frozen meat which will be separated during heat treatment.

Chilled meat is more suitable for both coarse and fine sausages as the meat batter can be more readily maintained at the optimal temperature. Another advantage is that neither a frozen-meat storage facility (minimum -18°C) nor a frozen-meat slicer is necessary.

*Methods of addition of salt, additives and spices*. The aim of using salt and/or nitrite salt in manufacturing comminuted meat is to separate the maximum quantity of salt-soluble proteins from the muscle cells. Optimum salt concentration for total extraction is 5–6 percent, and the optimum temperature is 3–5°C. Salt must be sprinkled on the pre-ground meat in the bowl as soon as possible. By mixing meat and salt at the slowest speed, salt starts to extract proteins from broken muscle cells. Half the ice water (1:1) is added to speed this process and control the temperature rise.

Additives contribute to improving and intensifying some properties of meat proteins, especially water-holding, water-binding and emulsifying capacities. In order to achieve this they are sprinkled on the meat after the salt has completed its effects, but always before adding fatty tissues.

To achieve uniformly and well-flavoured comminuted meat products, spices must be properly sprinkled in the meat batter and are added after additives, but always before fatty tissue to avoid them sticking to it causing a non-uniform colour. All spices for sausage formulation should be weighed and mixed in a plastic bag before adding to the meat batter.

*Hygienic aspects of comminuted meat*. Comminuted meat is more highly perishable owing to the large surface area exposed to many sources of contamination, the availability of the meat juice, water and nutrients, and the distribution of surface flora during comminution. The flora count varies greatly, presumably reflecting the initial contamination. For example, the count in ground beef is usually less than in minced pork. High counts will arise during comminution if the temperature of the sausage mixture is not properly controlled.

It may however be preserved in different ways: adding salt and nitrite, smoking, heat treatment, and adequate storage. The preservative effect of salt is due to its capacity to lower water activity to a level that inhibits bacterial growth. For a longer shelf-life other preservative methods must be used such as the addition of nitrite which has a marked bacteriostatic effect.

If the sausage mixture is kept at a relatively high temperature (above 20°C) before it is filled into casings, vegetative bacteria will grow and cause deterioration. In order to kill most micro-organisms present, emulsion-type sausages must be heat-treated to an internal temperature of 72–75°C. The product shelf-life is thereby significantly extended.

The product must be quickly chilled after heat treatment, in order to prevent growth of any surviving flora. After the final inspection it is ready for sale. Finished products must not come into contact with the floor or any other contaminated materials, especially raw meat or natural casings.

### Hygienic Faults Committed During Meat Comminution

The basic objective of meat comminution is to manufacture sausages of standard quality with a desirable hygienic level, and an acceptable shelf-life. For this reason, during comminution all improper handling which can promote growth of micro-organisms must be eliminated. The meat products are not only highly perishable, but may also become the source of foodborne diseases.

### Sources of Contamination During Comminution

*Unclean machines*. Meat batter and sausage mixture represent very good nutritive media for rapid growth of micro-organisms, owing to the presence of the soluble nutrients. If any interruption in production is greater than half an hour the machine must be emptied, the sausage mixture placed in the cooler, and the machine immediately washed. Failure to do so will lead to heavy contamination of the next batch. The contamination will be even more serious if residual material stays in unclean machines overnight or even during the weekend. To eliminate these hazards the machine must be washed immediately after use with hot water. The best method of washing for machines, containers and walls and floors is with a special cleaning machine. When attached to the water supply the machine can increase the water pressure tenfold to make cleaning very effective. Liquid disinfectant may also be added. Hot water (minimum 83°C) gives the best results.

During dismantling of any machine, either for daily washing or routine control of cleanliness, dismantled pieces must not be placed on the floor. They should be put in a plastic (not metallic) tray or vat and after washing hung on a rack, for draining overnight. The next day their cleanliness and sharpness must be checked before use. Water remaining after cleaning must be eliminated before leaving the workshop.

*Unhygienic handling*. There are many manual operations in a nonautomated production line for comminuted sausages. Unhygienic handling can provoke rapid bacterial growth, which will persist even after hightemperature heat treatment, resulting in a shorter shelf-life.

Meat trays, other tools, plastic aprons etc. must never be washed on the floor. Any material falling on the floor must not be reused in further processing. All dirty and contaminated material must be placed in a special watertight non-corrosive bucket, with a lid fastener, to prevent access to unauthorized persons and clearly marked in large red letters "condemned".

### Meat Extenders

Meat extenders are usually protein additives, defined as non-meat proteins. A wide variety of meat extenders are available for use in emulsion-type sausages to improve consistency and emulsifying and water-binding capacities. They can also serve to enhance protein content, improve processing yields and reduce formulation costs. The most important meat extenders are soy proteins, milk proteins, starch, flours and yeast. The permitted maximum amount of meat extenders in sausage production is usually 3.5 percent and is strictly regulated by law in certain countries.

### Soy Proteins

These are table proteins from soybeans in the form of flours, grits, concentrates and isolates, texturized, untexturized and extruded.

*Soy flour* is a screened, graded product, obtained after extracting most of the oil from dehulled soybeans.

*Soy grits* are particles of larger size, described as coarse, medium and fine.

Soy flours and grits may be defatted, low-fat, or high-fat. Owing to their functional properties (emulsion stabilization, and fat- and juice-binding), soybean flour and grits are used in the production of all emulsion-type sausages, meat patties and canned-meat products. However, both flour and grits give a slightly bitter taste to meat products, which limits their use (up to 2 percent).

*Soy protein concentrates* are high-protein products (not less than 70 percent). They possess good water-absorption abilities and are used in emulsion-type sausages, luncheon loaves and meat patties (2.5–3 percent).

*Soy protein isolates* are the most refined form of soy proteins, without water-soluble sugars and other insoluble components. Owing to their highprotein content (minimum 97 percent) they have excellent dispersing, emulsifying, gelling and water-and fat-binding properties. They are used in production of bologna, miscellaneous sausages, canned and dietary meat products.

*Textured soy proteins* are available in a variety of shapes, sizes, flavours (beef, pork, ham, chicken etc.), textures and colours. They are used as extenders in comminuted sausages, meat patties and canned-meat products. Some physical properties and typical chemical composition of different soy protein products are given.

## Application of Soy Proteins

Soy proteins and meat proteins interact differently with salt. While salt aids the extraction of salt-soluble proteins from meat, it has the opposite effect on soy proteins since it prevents their hydration. It is important therefore to add meat, soy proteins and salt in the correct order. Salt should be added only after hydration of soy proteins has been completed. For complete hydration sufficient water must be available and enough time allowed. The soy product/water ratio should be 1:4 or 1:5, with a minimum of three minutes' chopping with the knives and cutter at the highest speed. Soy proteins may be used in sausage production in dry gel and emulsion forms.

*Dry form.* Meat is placed in the bowl at slow speed and soy product is added. The correct amount of water is added at the highest speed to allow complete hydration of the soy proteins. When the mix is homogenous, phosphate is added and finally nitrite salt.

***Table:*** *Typical chemical composition of different soy products*

| ***Chemical components*** | ***Soy flour*** | ***Grits*** | ***Soy protein concentrate*** | ***Isolated soy protein*** |
|---|---|---|---|---|
| | (fine) | (coarse) | (coarse) | (fine) |
| | | | *(%)* | |
| Water | 6.5 | 8.0 | 9.1 | 4.8 |
| Protein | 53.0 | 52.0 | 65.3 | 92.0 |
| Sugar | 31.0 | 31.0 | 16.9 | - |
| Ash | 6.0 | 6.0 | 4.8 | 4.0 |
| Fibre | 2.5 | 2.5 | 3.6 | 0.25 |

*Gel.* Using soy gel ensures that the soy proteins are completely hydrated. To 15 kg of dry soy product, 60 kg of water is added. Complete hydration can be seen by the appearance of the gel consistency. Such gel, if used in sausage production, must be added to meat batter before fatty tissues and other additives. The remaining gel is stored in a cooler and, if not required until the next day, 2 percent salt is added.

*Emulsion.* Soy products can be added to emulsion-type sausages in the form of a pre-processed soy emulsion. These can be produced by processing with cold or hot water, depending on whether fatty tissues are used. The ratio between soy product, fatty tissues and water is 1:5:5 for firm and 1:4:4 for soft fatty tissues. In both cases, soy product is placed in a bowl, water is added, and the mix is comminuted to complete hydration (minimum three minutes). Fatty tissue is added and chopping is continued at the highest speed. For better bacterial stability 2 percent salt must be added. Coldprocessed emulsion can be used the same day, but hot-processed emulsion must be chilled overnight in flat trays in layers not more than 10 cm thick.

### Milk Proteins

These are casein, lactalbumin, lactoglobulin and other serum proteins. Casein is one of the most important proteins, used as an emulsifying agent in sausage manufacturing as caseinate (dry milk-protein isolate). It is a fine white powder with neutral taste and a protein content of approximately 94 percent. Unlike the coagulable soy proteins, egg albumen and blood plasma, milk protein does not coagulate, shrink or form a gel while heating.

If caseinate is added to meat before salt addition, it will surround free fat particles during comminution. Thus, the binding capacity of meat saltsoluble proteins may be enhanced to form a more stable emulsion.

In sausage manufacture, caseinate may be used in three ways:

*Prefabricated caseinate emulsion*. When making this emulsion, the order of addition, the necessary emulsifying time and temperature, and the protein:fat:water ratio are of major importance.

After the fatty tissue has been comminuted to a fine paste, dry caseinate must be added immediately, followed by all the hot water. After four to six minutes of comminution the emulsion is ready. During the last few rotations 1.5 percent salt is added. For high-emulsion stability it is necessary to reach a minimum emulsifying temperature of 45°C for pork and 50°C for beef fatty tissues. The emulsion must be chilled overnight in flat trays. Next day the emulsion may be added to the meat batter, together with fatty tissues.

The usual caseinate:fat:water ratio for the lower-viscosity caseinate (EM-6) is 1:5:5 and for high viscosity (HV) is 1:8:8. If cooked pork skin is used, the ratio of caseinate:fat:water and skin is 1:8:8:2 for sausages.

*Dry powder* caseinate may be used when prefabricated emulsion or gel are unavailable. As this is absorbed at the fat-water interface it is absolutely necessary to add caseinate before fatty tissues to ensure that the emulsion is heat stable.

*Prefabricated gel* is a concentrated solution of caseinate in water. Optimum results can only be expected when the caseinate is completely dissolved in water (1:8). It is recommended to moisten caseinate in the cutter with an equal amount of ice (1:1). After a few minutes of comminution the remaining seven parts of water are added. The gel is used in sausage formulations containing relatively small quantities of fat and in coarse comminuted sausages.

### Starch

This is a complex sugar of plant origin. The basic technological function of starch is to absorb released water and juice during heat treatment. The most frequently used starches are wheat, maize, potato, rice and manioc. Starch is used in doses up to 4 percent. During heat treatment of sausages, added starch binds part of the free water and swells, thus decreasing weight losses.

### *Flour*

Wheat flour is the most commonly used binder following hydration. Approximate chemical composition is starch 65–75 percent, gluten 8–14 percent, and water 12–16 percent. If flour is added in sausage formulation, care must be taken to prevent quick decomposition of the sausage mixture. Owing to the activity of flour enzymes, flour is commonly used in lowquality sausages (up to 4 percent).

### *Yeast*

Brewer's inactivated yeast with the bitter taste removed is frequently used. It is a fine yellow-grey or grey-pink powder, with distinctive taste and smell. Approximate chemical composition of yeast is: 53 percent proteins, 36 percent sugars and 3 percent water.

In the meat industry the most important application of yeast is based on the ability of its proteins to emulsify fat. A heat-stable emulsion is obtained if the yeast, water and fat ratio is not more than 1:5:5. In comparison with soy isolate and caseinate, yeast proteins have a lower ability to emulsify tallow. A heat-stable emulsion with tallow is only obtained with a ratio of 1:1.3:1.3.

Yeast can be used as an emulsifier in cooked and in canned chopped meat products and can easily replace caseinate in paste production. Yeast used in these products may improve their flavour. The content of free amino-acids improves the flavour of the meat product. Yeast also intensifies the flavour of certain seasonings, especially those with hot components.

### *Yeast Extract*

This is obtained from special yeasts cultivated on cane-sugar nutrient media. It contains many proteins, amino-acids and B-complex vitamins. It is used for improving the flavour of canned meat.

## CASINGS

Casings are special cylindrical containers used to protect sausages and various meat products. Since sausages are comminuted products they must be placed in some type of forming device to give them shape, to hold them together during further processing and for protection. Casings may be natural or artificial.

### *Natural Casings*

These are derived almost exclusively from the gastro-intestinal tracts of swine, cattle, sheep, goats and horses. Hog casings are prepared from stomach, small intestine (smalls), large intestine (middles) and terminal end (colon) of the large intestine (bung). Beef casings may be from the oesophagus (weasands), small intestine (rounds), large intestine (middles), bung and bladder. The intestines of sheep and goats are used to produce casings primarily for fresh sausages, frankfurters, hot dogs etc. Natural casings can be classified as narrow, medium and wide.

***Table:*** *Different diametres of small intestines*

| Type of animal | Diametre | | |
|---|---|---|---|
| | ***Narrow*** | ***Medium*** | ***Wide*** |
| | | *(mm)* | |
| Sheep | 16–28 | 20–22 | 22–24 |
| Hog | 30–34 | 35–38 | 38–40 |
| Beef | 34–37 | 40–43 | 43–46 |

*Treatment and storage.* Intestines intended for use as casings for sausages must be immediately processed after evisceration. First they are emptied and well flushed. Pig, sheep and goat small intestines are scraped thoroughly without inversion to remove the exterior (serous) and interior parts (mucous membrane), preserving the middle elastic muscle. The same method is applied to the large intestines of all animals.

Well-scraped, flushed and drained intestines are well sprinkled with common salt (dry salting) or saturated salt solution (wet salting), spun and tied into intestine completes. They are arranged into plastic casks. The top of the intestines is pressed down and salted again in order to eliminate air contact. These casks must be firmly closed and stored in a dark cooler. If intestines are prepared for long storage (more than three months), they must be salted with a mixture of salt and 0.2 percent antioxidant.

Large intestines and bladders can be preserved by means of air-drying. In this case, after scraping, flushing and a short draining intestines are blown up in order to eliminate the remaining water and tied at both ends in hanks, for quicker air-drying. They should be hung in a sunny, draughty place. When well dried they are sorted according to diametre and packed. Such dry casings do not require storage in a cooler.

*Application.* All salted natural casings should be thoroughly flushed inside with running water the morning before use. They are then dipped into warm water to regenerate their elasticity (small intestines 10–20 minutes and large intestines 30–60 minutes) and drained for a short time before use.

### Artificial Casings

The advantages of these are uniform cylindrical shape, a range of specific diametres, suitable tensile strength, resistance to damage, ease of use, variety of sizes and low microbial level. They can be filled uniformly and linked either by hand or machine into regular lengths.

Four types of artificial casings are available:

*Cellulose casings* are prepared from cotton linters in sizes ranging from 1.5 to 15 cm. They are easy to handle and fill, possess a high degree of resistance and are permeable to smoke when moist. The degree of permeability decreases as the casing dries during processing. Small cellulose casings are used for skinless frankfurters or wieners, skinless smoked link sausages and many other small sausage products (cocktail sausages). Small cellulose casings are also available in shirred form.

Large cellulose casings are available in three types: regular, highstretched and large. They are used in making all types of sausages and smoked meat. Cellulose casings are inedible and must be removed before consumption.

*Inedible and edible collagen casings* are regenerated from collagen extracted from skins and hides. The inedible collagen casings combine some of the advantages of both natural and artificial casings, especially their strength, uniformity and shrink characteristics. They must be removed prior to consumption. Edible collagen casings are mostly used for fresh pork sausages and frankfurters.

*Synthetic casings and bags* are impermeable to smoke and moisture. They are used with unsmoked products such as liver sausages or products which are heat-processed in water or a retort.

### Tying of Casings

Artificial casings are delivered either tied at one end with loops prepared in the factory or in hanks. As casings in hanks can be easily stored and their length and the method of tying can be chosen at will, they are well suited to small-scale production. One end of the casing is submerged in cold water (one to two minutes), then tied either in a "doll" or "bow" knot. To make a "doll" knot, place a string on the casing and twist around, slip the thumb and forefinger on the string 1 cm towards the end of the casing, pass the string through the bight and set the knot, tighten the string, and prepare the loop and tighten with a double knot.

When tying a "bow" knot, casings need not be submerged except the end of every dry casing. The method is to place string on casing and tighten by means of a simple knot, fan-shape casing end with knot in the middle, and tie knot underneath the wing (the tied casing has no loop). This method of tying prevents slipping off.

### Hamburger-type Products

Hamburger-type products are made form minced meat and fatty tissue to which salt and seasonings are added. Soy is a common additive but curing agents need not be used.

### Raw Material

Well-chilled Meat I and Meat II and fatty trimmings are the basic raw materials. All beef, all mutton, all pork or combinations of different meat and fatty tissues in various proportions may be used.

### Additives

Extruded soy proteins (ESP) are often used in manufacturing hamburgertype products. They should be hydrated in water (70°C for 35–40 minutes) in the ratio of ESP to water of 1:1, 1:2 or 1:3. Salt, seasonings and dry or fresh onions are the other basic ingredients. If these products are intended for long-term storage some antioxidants may be added.

***Table:*** *Common composition of hamburger-type products*

| | Products | | |
|---|---|---|---|
| ***Raw material*** | ***All-beef*** | ***Beef-pork*** | ***All-pork*** |
| | | *(%)* | |
| Beef I | 53 | 25 | - |
| Beef fatty trimmings | 10 | 5 | - |
| Pork I | - | 40 | 65 |
| Pork fatty trimmings | - | 10 | 20 |
| ESP: water (1:1) | 30 | 15.5 | 10 |
| Breadcrumbs | 2.5 | - | 0.5 |
| Salt | 2 | 2 | 2 |
| Onions | 2 | 2 | 2 |
| Seasonings (according to taste) | 0.5 | 0.5 | 0.5 |

### *Manufacturing*

Selected meat is minced through a 5-mm, fatty trimmings through a 3-mm, and hydrated ESP through a 3-mm plate. After mincing, the components are mixed with the additives and seasonings until all components are evenly blended. The mixed batter is then ready for forming and shaping. Weight and shape can vary but patties are usually round, weighing 80–120 g and 5–10 mm thick. After freezing, the hamburgers are packed into suitable plastic bags or cardboard boxes, and immediately transported to a storage room.

Beef II or Pork II are used with lower-quality products and ESP is hydrated with more water (1:2, 1:3). Hamburger-type products should contain 10 to 25 percent fat but if hydrated ESP is used up to 30 percent may be added.

### *Storage and Shelf-life*

Hamburger-type products are stored at -20°C to give a shelf-life of 90 days. They may be cooked without thawing by grilling, frying in deep fat or toasting on a hot plate.

# 7

# Dry Sausages

Dry sausages are specific coarse-comminuted meat products whose successful manufacture depends upon bacterial fermentation. At some stages of processing, usually during smoking, these sausages are deliberately held at temperatures which encourage bacterial growth and fermentation. Dried sausages should never exceed 30°C at any stage as this would stop growth. The useful organisms responsible for desirable fermentation (lactic acid-producing bacteria) originate from the natural flora of the meat, processing equipment and the plant environment. Fermentation causes a characteristic tangy flavour to develop, resulting from the accumulation of lactic acid and many other fermentation compounds. The pH usually falls to 4.8–5.4. Another basic processing preservative step is dehydration achieved by keeping the product under controlled temperature and air humidity (drying and ripening).

One of the distinctive features of dry sausages is that they are processed uncooked. The low pH (high acidity), low-water content and high-salt content extend their shelf-life. Some but not all dry sausages are smoked. The best known are dry pork sausages, dry beef sausages, mixed dry sausages, summer sausages and salamis.

## Raw Material

In general, dry sausages are composed of two-thirds meat and one-third fatty tissues to which curing agents and spices are added. Meat I and Meat II of all species of slaughter animals can be used, including camel, donkey and horse meat, but rarely mutton, goat or venison. Trimmings are not used owing to their softness, neither are shanks or head or boar meat. Pork jowl and back fat, beef external fatty tissues and humps are used as fatty components.

Only well-chilled (-1° to 0°C) and/or frozen (-1° to -18°C) meat is used. Frozen meat is thawed for 48 hours on slanting racks to allow the drip to run off.

The meat is chopped with sharp regular cuts, without crunching. The temperature of meat and fatty tissues should be adjusted to stay in the range of -2° to 5°C during chopping so that temperatures in the filler will be between-1° and -3°C.

If the temperature in the filler is higher, drops of fat are deposited on the interior walls of the filler horn. During further filling they are pressed into the casing and will grease the interior of the casings lowering their porosity, making smoking, drying and ripening more difficult.

### Main Additives

*Salt* (28–37 g per kg) is used to prevent the growth of many undesirable aerobes, favouring the growth of non-spoilage halophile and halotolerant bacteria. Salt also extracts the salt-soluble proteins to form a protein gel which binds the pieces of meat and improves sliceability of the finished product.

*Curing substances* are used in the form of nitrite salt or a dry mix of common salt with 0.6 percent nitrite salt and nitrate (0.3–0.5 g nitrate per kg).

*Sugars* speed acidification and are transformed into lactic acid by certain bacteria. Dextrose is the most commonly used sugar (8–10 g per kg) but can be replaced by saccharose.

*Spices.* White and black pepper (0.5–3 g per kg), ground or crushed, are the most frequently used spices. Fresh crushed garlic, paprika, cardamom, mustard etc. are also used. In some countries wine is added to improve the flavour. Some antioxidative substances may also be added.

*Starter cultures.* To overcome the problems associated with bacterial fermentation, starter cultures of selected lactic acid-producing flora have recently been used.

The starter culture provides a predominant flora of the desired bacteria (*Micrococcus, Pediococcus cerevisiae* etc.) in the sausage mixture, and fermentation is initiated within a minimum time.

***Table:*** *Typical composition of some dry sausages and salami*

| Components | Variation of dry sausages | | | All-pork dry sausage | All-beef dry sausage | Beef soujouk | Salami |
|---|---|---|---|---|---|---|---|
| | 1 | 2 | 3 | | | | |
| | | | | (%) | | | |
| Pork I and II | - | 29 | 30 | 69 | - | - | 67 |
| Beef I and II | 61 | 47 | 37 | - | 74 | 86 | - |
| Mutton I and II | - | - | 7 | - | - | - | - |
| Pork back fat | 14 | 19 | 10 | 10 | - | - | 28 |
| Pork jowl | - | - | 5 | 5 | - | - | - |
| Side pork fat | 20 | - | - | 10 | - | - | - |
| Beef ext. fat | - | - | - | - | 10.5 | 5 | - |
| Hump (zebu) | - | - | 4 | - | 10 | 5 | - |
| Salt | 3.5 | 3.5 | 3.5 | 3.5 | 3.5 | 3.5 | 3.5 |
| Nitrate | 0.05 | 0.05 | 0.05 | 0.05 | 0.05 | 0.05 | 0.05 |
| Sugar | 0.6 | 0.6 | 1.0 | 0.45 | 0.4 | - | 0.5 |
| Garlic | 0.25 | 0.25 | 0.45 | 0.5 | 0.45 | 0.2 | 0.3 |
| White pepper | 0.2 | 0.2 | 0.3 | 0.3 | 0.3 | - | 0.3 |
| Black pepper | 0.1 | 0.1 | 0.2 | 0.2 | 0.2 | 0.25 | 0.1 |
| Paprika | - | - | 0.2 | - | 0.3 | - | 0.25 |
| Anise | - | - | 0.1 | - | 0.1 | - | - |
| Mustard | - | - | 0.2 | - | 0.2 | - | - |
| Wine | - | - | 1.0 | 1.0 | - | - | - |
| Cognac | 0.3 | - | - | - | - | - | - |
| Rum | - | 0.3 | - | - | - | - | - |

### Casings

These may be natural or synthetic. Natural casings are usually small intestine of hog (for small-diametre dry sausages), beef small intestine (for middle-diametre dry sausages), and large hog or beef intestine (for largediametre dry sausages).

As for salami (diametre more than 40 mm) and summer sausages the most convenient casings are small horse intestine. The best synthetic casings are the so-called “dry sausage fibrous casings”, as they adhere very well to the product as it shrinks during drying.

***Table:*** *The most important parametres in manufacturing dry sausages*

| ***Technological operations*** | | ***Parameters*** |
|---|---|---|
| Choice of raw materials | meat<br>pork<br>beef<br>fatty tissues | -1° to -30°C<br>pH 5.8 and lower<br>pH 6.0 and lower<br>-10° to -30°C |
| Cutting | cutting room | not more than +15°C |
| Chopping | finished sausage<br>mixture | -2°C to -5°C<br>pH 5.9 and lower<br>$a_w$ 0.97–0.96 |
| Filling | filling | -1° to -3°C |
| Curing | curing room<br>(2–5 days) | 21° to 24°C<br>RH 75–80% |
| Smoking | dry sausages | maximum 30°C |
| Drying and ripening: | drying room<br>(2–4 days) | 18° to 25°C, RH 94–90;<br>air velocity 0.5–0.8 m/sec |
| first stage | product | pH 5.6–5.2<br>$a_w$ 0.96–0.94 |
| | drying room<br>(5–10 days) | 18° to 22°C, RH 90–80%;<br>air velocity 0.2–0.5 m/sec |
| second stage | product | pH 5.2–4.8<br>$a_w$ 0.95–0.90 |
| | drying room<br>(to 90 days) | 12° to 15°C, RH 80–65%;<br>air velocity 0.05–0.1 m/sec |
| third stage | product | pH 5.3–5.8<br>$a_w$ 0.92–0.85 |
| Storage | store-room | 10° to 15°C, RH 80–65%;<br>air velocity 0.05–0.1 m/sec |
| Package | packing room | 10° to 15°C, RH 65–75% |

## Composition of Some Popular Dry Sausages

Like all other sausages there is much variation in the composition of dry sausages and salamis. A common factor is that they do not contain any added cereal, ice or water.

## Manufacture

In the successful manufacture of many varieties of dry sausages and salami, a great deal of technical skill is indispensable. In manufacturing coarsely chopped sausages the grinder and mixer are used but for less coarsely chopped sausages only the cutter is used.

In the first case, meat and fatty tissues pass through the grinder (beef 15-mm, pork 2–6-mm and fatty tissues 8–10-mm plate) and the sausage mixture is made in the mixer. In the second case, chopping and mixing are done in the cutter at slow speed. After adding the curing agents the meat is chopped to the required size then fatty tissues and all other components are added. The mixture is firmer and without residual air if a vacuum is applied. The sausage mixture must be firmly filled into casings.

The casings are punctured with small needles over the entire surface to allow entrapped air to escape. Tied sausages are hung on sticks and transferred into a curing room. Curing time depends on the sausage diametre, three days for diametres up to 3 cm, and five days for larger. During this period, the cured meat colour is developed and fermentation is initiated.

After curing, sausages are transferred to a drying room. The rate of drying is controlled by keeping the products between narrow limits of both temperature and RH. Too rapid drying will result in the formation of an outer crust on the sausages, which will retard or stop internal drying. If the drying rate is too slow and RH is too high, then surface mould, yeast and bacterial growth are excessive. During drying and ripening, flavour develops, texture changes and the product hardens. Degree of drying can vary according to local preferences. Ripening can either follow or precede cold-smoking, depending on the particular product.

Dry sausages must be stored in an environment in which temperature and humidity do not provoke overdrying. They are ready for sale immediately after production. All dry sausages may be sliced and packed under vacuum in different consumer sizes. They are consumed sliced, as starters or in sandwiches.

Only complete fulfilment of all technological parametres at all stages of production can guarantee the desired quality of finished products. Dry sausages, like dry hams, are high-quality meat products. The main characteristics of dry sausages are agreeable bouquet, i.e. flavour and taste of a well-matured cured-meat product, attractive colour, good sliceability and long shelf-life.

In order to achieve all these, sufficient time must be allowed of at least a month for small-diametre sausages and three to six months for large-diametre sausages and salami. Organoleptic tests and weight-loss control may confirm the end of ripening. Water content of finished dry sausages (of any diametre) should be between 25 and 30 percent. Sausages should not be overdried or they are hard to chew and less acceptable to consumers.

## Semi-dry Sausages

They are produced by quick drying without ripening. Depending upon the diametre, this lasts from two to three weeks. In order to shorten the process, reducing agents (such as Glucono-delta-Lactone) and a starter culture must be added to the meat mixture. Weight loss is lower than in dry sausages, so the water content of the finished product is always greater than 40 percent, resulting in a shorter shelf-life, a sour taste and a poor flavour. Sliceability is also owing to reduced binding of meat and fatty tissues. Semi-dry sausages have a maximum shelf-life of one month at room temperature.

## Frequent Faults Committed During Production of Dry Sausages

*Creases or detachment* appear in sausages with synthetic casings if the filling dries at a faster rate than the casing. Small creases become larger as drying continues, and overdrying causes detachment of the casing. This can be prevented by using more elastic casings and by a slower drying rate.

*Crust formation* results from rapid drying, especially at the initial stage, and can be detected by careful palpation. If the crust is not firm a preventive measure is to increase the humidity in the drying room to soften and rehydrate the sausage.

*Greasy casing* results from using soft fatty tissues or from overheating the sausage mixture. Melted fat, under pressure during filling, greases the interior walls of the casing reducing its porosity and preventing normal water migration from inside the sausage. The consequence is a soft sour sausage. If the melted fat penetrates the casing walls, then the product also becomes greasy.

*Sausage mould*, usually greenish in appearance, is the sign of contaminated equipment and workshop. It is eliminated by using good antifungicidal agents in cleaning operations.

*Sausage sliminess* is the result of heavy micro-organism growth on the surface of the sausage casing, encouraged by too high a temperature and air humidity. Partial sliminess can be removed by washing in salty water, followed by dripping and more intensive and dense smoking. This fault may appear when the product is stored in cardboard boxes for a long time.

*Sour sausage*, being invisible, is unfortunately usually detected only at the end of drying during the final control. It has a sour taste, semi-rigid consistency and the periphery is darker than the centre.

It is the result of the intensive growth of lacto-acid bacteria, feeding on the added sugar. Preventive measures are to decrease the added sugar and increase the added salt. Such products must be properly dried and ripened.

*Sausage porosity* usually appears in large-diametre sausages. Air in pores causes decomposition of fat, leading to changes in colour and taste.

*Poor sliceability* is the result of insufficient curing and salting, uneven distribution of curing agents, and/or short curing. The lack of salt-soluble proteins reduces binding between meat and fat particles.

*Blown-up sausages* due to gas produced by sudden micro-organism growth result from contamination of raw materials, equipment, tools and unhygienic production.

Dry sausages are a very nutritive meat product which may be eaten immediately after production, without any additional heat treatment. They are sold entire or packed in pieces in plastic bags. All dry sausages can be sliced and packed under vacuum, in various consumer sizes. Dry sausages are usually covered in very thin slices in sandwiches, or eaten as starters.

## Cooked Sausages

Cooked sausages are fine-comminuted sausages representing an important group of meat products. The basic technological aim is to increase the natural WHC of the meat batter to bind considerable quantities of added water. Cooked sausages may be made of uncooked or cooked raw materials.

### *Cooked Sausages Made of Uncooked Raw Materials*

These are emulsion-type sausages. The finished products are heat-treated to between 75° and 80°C to achieve desirable organoleptic properties and bacterial stability. They are juicy, tasty, easily chewable and digestible. To achieve such juiciness, about 25 percent water is added to the meat during manufacturing. Phosphate must be added to improve the WHC of the meat proteins. Specific amounts of fatty tissues must also be added to achieve the characteristic consistency and flavour.

*Raw material.* Many different kinds and qualities of meat, organs and fatty tissues can be used. Beef, veal and pork are more frequently used, but camel, mutton, goat and venison can also be incorporated. Practical experience has shown that more desirable cooked sausages of the emulsion type are manufactured when hot-boned meat is used. In this case phosphate is not necessary, because hot-boned meat has a high WHC in the first three to four hours after slaughter and can emulsify 22 percent more fat than chilled or frozen. If hot-boned meat is not available only well-chilled and/ or frozen meat should be used. The chemical composition, water-binding and fat-emulsifying properties of the raw materials must be known in order to make products of uniform quality.

Offals such as heart, tongue, spleen, lungs, stomach, tripe or non-lactating udder are not commonly used in the manufacture of emulsion-type sausages. However, special products of this type may contain smaller quantities of certain offals.

Fatty tissues are also important raw materials. They contribute to the palatability, tenderness and juiciness of sausages. Variation in the ability to emulsify fat is due to the amount of soluble proteins potentially available, and their emulsifying capacity.

***Table:*** *Raw materials suitable for manufacturing cooked sausages made of uncooked material*

| ***Cattle*** | ***Pigs*** |
|---|---|
| Boneless primal cuts (Meat I) | Boneless primal cuts (Meat I) |
| Boneless bull and cow meat (Meat I & II) | Trimmings from primal cuts (Meat II) |
| Beef trimmings | Trimmings from blade and neck bones |
| Head and cheek meat | Head and cheek meat |
| Diaphragm meat | Diaphragm meat |
| Blood plasma | Skin |
| Suet and brisket fat | Blood plasma |
| | Jowl, backfat |
| Calves | Sheep and goats |
| Boneless primal cuts | Boneless primal cuts |
| Cheek meat and trimmings | Cheek meat and trimmings |

Ice water is used as a carrier for the curing agents, and improves the extraction of meat protein and the hydration of meat-extender proteins. The amount of added ice water depends on the raw material used. Water can be partly substituted by blood plasma which is an additional source of protein and improves the water-binding and emulsifying properties of the product. Regulations for emulsion-type sausages in many countries limit fat content to a maximum of 30 percent and water content to 40–60 percent in finished products.

*Additives.* A variety of additives can be used in addition to meat extenders. Phosphates and seasonings (spices, sweetners, glutamates) may be used to create a range of products. If frozen and chilled meat is used, 0.3–0.5 percent phosphate must be added.

Salt and pepper form the basis for a cookedsausage seasoning mixture. All other seasonings are only supplementary to them. Ascorbic acid, sodium ascorbate or sodium erythorbate (about 0.4 g per kg) are used in a 10-percent water solution at the end of the comminution.

*Casings.* Natural or synthetic casings can be used but in both instances they must be firmly filled.

*Manufacture.* The basic aim in cooked-sausage manufacturing is to make a heat-stable meat batter, a meat "emulsion", consisting mainly of water and protein. Heat treatment transforms it from a viscous form to a rigid and elastic solid structure which can be considered as a protein gel with entrapped fat particles. If the protein gel is strong enough, the fat will not be separated during heat treatment.

*Principles of mincing.* Mincing completely destroys the meat structure. Meat proteins in the presence of salt, phosphates and cold water are dissolved forming a system consisting of a solution of salt-soluble proteins and muscle and connective tissue particles. When fatty tissue is added the fat particles will be emulsified with the salt-soluble proteins during comminution. The proteins will therefore be able to cover the total fat surface, surrounding each fat particle and so stabilize the emulsion.

During comminution the structured breakdown of meat and fatty tissues occurs and new systems are formed: minced meat, suspension, after addition of salt, phosphate and ice water and emulsion, after addition of fatty tissues and stable sausage mixture after addition of all other additives.

*Principles of filling.* Cellulose and collagen casings are ready for transferring directly to the filling horn. Salted, natural casings should be flushed thoroughly prior to use, and filled firmly.

*Principles of heat treatment.* Proteins show a great tendency to bind to and cross-link with each other upon heating. The binding properties of comminuted sausages originate from the heat-induced gelation of meat proteins. Gelation begins at about 43°C and produces a three-dimensional network structure which stabilizes the fat and water in comminuted meat products and binds meat and fat pieces together.

***Table:*** *Casings used in manufacturing cooked sausages made of uncooked raw materials*

| ***Type of sausage*** | ***Length (cm)*** | ***Diameter*** | ***Casings*** |
|---|---|---|---|
| Frankfurter | 12–14 | small | sheep, goat and hog casings, or collagen or cellulose casings of similar diameter |
| Vienna sausage | 12–14 | small | as above |
| Hot dog | 16–18 | small | as above |
| Knackwurst | 12–14 | small | as above |
| Bologna | | | |
| long bologna | 40–100 | medium | beef middles or weasands, or cellulose casings of similar diameters |
| large bologna | 20–50 | large | beef bungs or cellulose casings of similar diameter |
| ring bologna | 40–50 | large | beef rounds or cellulose casings of similar diameter |
| Veal sausage | 40–60 | large | beef bungs or cellulose casings of similar size |

*Frequent faults committed during production*

- Separation of fat in the form of fat caps or fat pockets develops during heat treatment, due to emulsion breakdown, as a result of improper formulation (too much fatty tissue, especially soft or insufficient emulsifier), incorrect order of addition of raw materials and additives, and overcooking.
- Holes in sausage mixture caused by improper filling.
- Shrinkage of cooked sausages caused by improper smoking and/or heat treatment (temperature too high and RH too low).
- Difficult peeling of casings, can be the result of too much connective tissue in the formulation.
- Uneven colour of smoked sausage surface (unsmoked spots) occurs at places where links touch each other.
- High porosity of sausage mixture due to entrapped air.

***Table:*** *Typical composition of different types of cooked sausage made of uncooked raw materials*

| ***Ingredients*** | ***Beef frankfurter*** | | ***Mixed frankfurter*** | | ***Pork frankfurter*** | ***Vienna*** | ***Bologna*** |
|---|---|---|---|---|---|---|---|
| | 1 | 2 | 1 | 2 | | | |
| | | | | *(%)* | | | |
| Meats | | | | | | | |
| Beef I and II | 35.0 | 30.0 | 25.0 | 20.0 | - | 30.0 | 25.0 |
| Pork II | - | - | 20.0 | 20.0 | 35.0 | 20.0 | |
| Beef fatty trimmings | 33.0 | 20.0 | - | 4.0 | - | - | - |
| Pork fatty trimmings | - | - | 5.0 | 5.0 | 5.0 | 42 | 4 |
| Beef fatty tissues | 6.0 | 10.0 | - | - | - | - | - |
| Other by-products | - | - | 7.0 | 8.0 | - | - | - |
| | | | | | | | |
| Other ingredients | | | | | | | |
| Ice water | 21.0 | 16.0 | 23.0 | 22.0 | 25.0 | 28 | 24 |
| Meat extenders | 2.0 | 3.0 | 3.5 | 3.0 | 3.0 | 3.5 | 3 |
| Phosphates | 0.3 | 0.4 | 0.4 | 0.3 | 0.3 | 0.5 | 0.5 |
| | | | | | | | |
| Curing ingredients | | | | | | | |
| Nitrite salt | 1.6 | 1.8 | 1.8 | 1.8 | 1.8 | 2.0 | 2.0 |
| Sugar | 0.5 | 0.2 | 0.4 | 0.2 | 0.3 | 0.6 | 0.3 |
| Glutamate | 0.15 | 0.2 | 0.2 | 0.2 | 0.2 | 0.2 | 0.2 |
| | | | | | | | |
| Seasonings | | | | | | | |
| Pepper | 0.12 | 0.13 | 0.3 | .2 | 0.2 | 0.4 | 0.35 |
| Coriander | 0.08 | 0.1 | 0.1 | 0.1 | | 0.3 | 0.15 |
| Nutmeg | - | - | 0.1 | 0.1 | 0.1 | - | - |
| Sage | 0.05 | 0.05 | - | 0.05 | - | - | - |
| Cinnamon | 0.05 | 0.05 | - | 0.05 | - | - | - |
| Fresh garlic | 0.05 | 0.02 | 0.1 | - | - | - | 0.35 |
| Clove | - | - | 0.05 | - | 0.05 | - | - |
| Mace | 0.1 | 0.05 | - | 0.1 | - | 0.2 | 0.15 |
| Ginger | - | - | 0.05 | - | 0.05 | 0.3 | - |

*Different products*. The best-known emulsion-type cooked sausages made of uncooked raw materials are frankfurter, Vienna sausage, hot dog, knackwurst, safalada, bologna and veal sausage. Frankfurter-type sausages are sold under many names: frankfurters, wieners (braided in groups of links), Vienna-style sausage (twisted into a chain of links), hot dog, knacker, etc. All these types normally contain more beef than pork (1.5:1).

The choice of raw material largely depends on the availability of the meat, the eating habits of the consumers and religious practices. In general, the lean-meat content varies between 25 and 40 percent, the average fat content is 25-30 percent, and 10-30 percent water is added. In low-quality frankfurters, 10 percent of the lean meat can be replaced by pork skin or organs. To maximize yield part of the fatty tissues and part of the water may be pre-emulsified (10–15 percent).

***Table:*** *The most important parametres in manufacturing emulsion-type cooked sausages made from uncooked raw materials*

| ***Technological operations*** | | ***Parameters*** |
|---|---|---|
| Choice of raw material | meat | warm meat (hot-boned), chilled +2°C, or frozen minimum - 18°C |
| | fatty tissues | -1° to +2°C |
| Comminuting | finished sausage mixture | 16° to 18°C (all pork)<br>18° to 20°C (all beef)<br>pH 5.7–6.2, $a_w$ 0.96–0.98 |
| Filling | filling mixture | max. 20°C (pork); 22°C (beef) |
| Smoking + heat treatment | smokehouse | 54°C (30 minutes), and 60 minutes; raise gradually to 80°C and hold 20–30 minutes |
| | product | 70° to 75°C, pH 5.8–6.3, $a_w$ 0.96–0.98 |
| Cooling | cooling room (12–14 hours) | -1° to +2°C<br>RH about 90% |
| | product | maximum temperature +5°C |
| Storage | storage room | 8° to 10°C<br>RH 80-85% |
| Package | packing room | 12° to 15°C<br>RH 70–75% |
| Sale | sale rack | t = 5°C |

If a high-speed cutter is not available, the first manufacturing step is grinding. The meat and trimmings must be ground through a 3-mm, and fatty tissue through a 5-mm plate.

The temperature of all raw material after grinding should not exceed 3°C. If a high-speed cutter is available, then a grinder is not necessary as the sausage mixture can be made in the cutter. Attention should be paid to the order in which raw materials are added (meat, curing agents, ice water, meat extenders, fatty tissues and seasonings). Water must be added early enough to ensure protein extraction. It is important to control the temperature during comminution, not only for the extraction of salt-soluble proteins but also for the hydration of meat-extender proteins. The final temperature of the sausage mixture should not exceed 18°C if pork fatty tissues are used, and 2°C if beef fatty tissues are used.

After filling in natural, synthetic, edible or inedible casings and linking, products should be smoked immediately. They are hung on sticks and placed on smokehouse racks. Links should be separated on the stick so as not to touch each other, otherwise smoking and colour development will not be uniform. Surfaces in contact with the stick or with other links do not dry out sufficiently, favouring the growth of micro-organisms forming slimes, especially in warm conditions. This can be eliminated by shifting the links slightly at some time during smoking to expose the surfaces which have been in contact with the stick. This can be minimized by using "T" profile metal sticks with apertures.

Smoking imparts an appetizing flavour and develops a desirable goldbrown smoked colour. It also contributes to the development of a smooth surface or skin beneath the cellulose casing, that facilitates peeling of the casing prior to packaging. Smoking lasts from one to three hours, depending on desirable colour, temperature and humidity, and type of smokehouse employed. A high smoke concentration and a rather high temperature are desirable during the final period of smoking. Smoking and heat processing start at 54–57°C and the temperature is gradually increased (5.5°C every 15 minutes), until it reaches 82°C. Smoking continues until an internal temperature of 70–75°C is reached. The next step is cold showering until an internal temperature of 32°C maximum is reached. The product is then placed in a cooler.

Vienna sausages are manufactured using beef and pork trimmings with a maximum quantity of water (28–30 percent) and meat extenders, tied in similar lengths to frankfurters (12–14 cm).

Hot dogs have the same formulation as Vienna sausages but are tied longer than any other emulsion-type small-diametre cooked sausages (16–18 cm).

Safaladas/knackwursts have a frankfurter composition but the structure is finer than any other cooked sausage. It is obtained by passing the sausage mixture twice through an emulsion mill, after being manufactured in the cutter.

Bologna sausage is a typical emulsion-type cooked sausage filled into large casings. Curing is as in frankfurters but less water is added. Long bologna is filled into beef middles or weasand, large bologna into beef bungs, and ring bologna into beef rounds. Cellulose casings of corresponding size may also be used. Large bologna must be supported either by wrapping it two or more times with twine and hanging it on a loop in the twine or by supporting it in a stockinette bag. Bologna requires longer smoking (three hours for long and five for large) and cooking than frankfurters.

All parts of the bologna must reach 71°C during cooking or the inside of the sausage may become discoloured before it is sold to the consumer. After cooking the bologna is chilled in cold water, superface dried, then placed in a cooler.

*Shelf-life* of cooked sausages is about four to five days, or up to ten days in retail packages (vacuum) held at 10–15°C or as high as 12 to 14 months in cans or jars.

Small and medium-diametre cooked sausages are reheated before eating in order to regain their firmness and juiciness. Reheating is done in boiling water for about ten minutes (or by toasting or grilling). Cooked sausages are eaten while hot with baked goods (rolls, buns), mustard, horseradish, ketchup etc. Large-diametre sausages are cut into slices as a cold starter or used in sandwiches, Russian salads and similar light meals.

### Cooked Sausages Made of Precooked Raw Material

These are sausages made from edible offal and lower-quality meat and have a specific flavour. The raw materials are precooked, minced, mixed, filled into casings and finally heat-treated. Fine-comminuted sausages are processed in the cutter and coarse-comminuted in the grinder. There are also sausages with a mixture of both fine- and coarse-comminuted pieces of meat, tongue, liver, fatty tissue (all these components are most often cured). Uncooked cured blood (2 percent nitrite salt) is often added and precooked pigskin is also very often used, except with all-beef products in which tendons and other connective tissues are used instead.

*Raw materials* are all edible offals, head meat, lower-quality meat, fatty tissues and pigskin. All components except pigskin are fully cooked. Pigskin is partially cooked until soft enough to be ground in the grinder. Its function is to give the finished product its typical firmness.

Precooking is done in boiling water with sufficient water in the cooking vats to cover all the material. Cooking time depends on the size of pieces. Fully cooked raw materials are comminuted without delay. If comminution is done in the cutter then no blending is necessary.

*Additives* used are nitrite salt, caseinate, blood-plasma powder, egg white, gelatine and seasonings. All except the seasonings can be added at a rate of up to 2 percent.

*Casings.* Natural and synthetic casings can be used but in countries with a hot climate, if continuous refrigerated storage during retail sale is not possible, priority should be given to synthetic casings. Such casings are also more convenient because they can be exposed to higher temperatures (up to 115°C retortable sausage) during heat treatment giving a longer shelf-life.

*Composition.* There are four types of precooked sausages which differ in composition and manufacturing method:

- liver paste;
- liver sausage;
- blood sausage; and
- cooked sausage with high collagen content.

In all formulations meat can be substituted with other edible offal. Liver paste is fine-comminuted with high spreadability.

Liver sausage contains 10–20 percent liver, usually coarse-ground, and other edible offals.

Blood sausage contains 10–20 percent whole blood with nitrite salt (not precooked). Other components are precooked meat, edible offals, fatty tissue and pigskin. Fatty tissue must be cooked sufficiently in order to separate fat with a low melting point. Components can be coarse- or fine-comminuted as in other precooked sausages. This type of sausage has a firm consistency due to swollen connective tissue components and gelatinized collagen.

Cooked sausage with high-collagen content resembles coarse-textured liver sausage but has more connective tissue (collagen and elastic of pigskin and beef tendons). Sometimes up to 10 percent dissolved gelatine solution is used to produce a sausage described as aspic in casing.

*Heat treatment.* The mixture is filled into the casings while still hot and the sausages are immediately heat-treated. Cooking time should be set according to the type (natural or synthetic) and diametre of casing. Natural casings exposed to temperatures higher than 85°C will burst, therefore their shelf-life is shorter (up to ten days at 5°C). Synthetic casings can be cooked in boiling water for about 90 minutes, or even in an autoclave at 110°C for 60 minutes. If they are kept at temperatures below 10°C they will have a shelf-life of one month.

*Frequent faults committed during production*

- Undercooking resulting in a shorter shelf-life
- Overcooking may result in an oversoft consistency
- Insufficient connective tissue will also result in a soft consistency, and the filling will fall apart
- Fat separation may occur in the liver paste if insufficient emulsifier is added.

The most serious fault is not to fill the casings immediately and not to heat-treat them immediately. In this case the product can spoil even before the heat treatment.

*Different products.* There are many products of this type which differ in the raw materials, grade of comminution and seasonings. Sometimes cooked rice is added to the sausage mixture.

*Eating method.* Liver paste is spread on bread or is used to make sandwiches. Other types of sausage are cut into slices and consumed cold.

## Simple Canning

The canning process involves two essential operations: the product must be heated at a sufficiently high temperature and for long enough to make it fully or commercially sterile, and it must be sealed in a hermetic container which will prevent recontamination of the product.

Commercial sterility differs from total sterility in that some organisms may survive the heat treatment but the conditions which prevail in the container during storage do not allow these to grow, produce toxins or spoil the product. However, in regions having a tropical climate, canners strive for total sterility of their canned products. The need to achieve at least commercial sterility determines the minimum heat process to be applied to a product. There is sometimes a temptation for canners to use less than the recommended minimum heat process. This may result in the product not being commercially sterile.

It may then become toxic and poison consumers, or the product may be spoilt and the cans may swell and have to be destroyed. It is essential for people dealing with canning to know what heat processes are required for their products, how these are to be applied, and the nature of the risk they take if less than minimum processes are used. This is why in numerous countries only people having achieved the required level of expertise in canning technology and heat process calculations are authorized to approve heat processes.

Commercial sterility is obtained in meat products which belong to lowacid foods (pH higher than 4.6) if the process applied is severe enough to inactivate the spores. Therefore, meat products are usually processed in steam (or water) under pressure at 116–121°C and sometimes in steam at 140°C. The organisms which are capable of spoiling meat products include those which form heat-resistant spores, thus high-temperature processes are needed to make them commercially sterile.

Usually heat processes for canned products are designed to inactivate large numbers of spores of the organism *Clostridium botulinum*. Although these spores are not as resistant as the spores of some other *Clostridium* and *Bacillus* types, *C. botulinum* is capable of producing lethal toxins, sometimes without swelling the container or obvious alteration of the appearance of the product. Since this organism presents a public health risk, recommended heat treatments must have a large safety margin.

The severity of heat processes for canned meat products is measured in terms of $F_0$-values which means that the product received a heat treatment with the same inactivating effect as exposure for one minute at 121°C. For example, one minute at 121°C gives the same amount of inactivation of spores as four minutes at 115°C or 13 minutes at 110°C or 40 minutes at 105°C, so all those processes will have the same $F_0$-value. $F_0$-value for the majority of canned meat products ranges between one and ten. Larger canned products require higher $F_0$-values, even up to 20–25, owing to the longer period needed for heat penetration.

## *Meat Products Suited for Canning*

Meat products made from chopped cured meat that can be canned include corned beef (consisting of large and small pieces of beef, blanched to give a lower water content); corned mutton (similar product made from mutton); beef hash (similar product made from lower-quality meat to which some edible offal is added); minced beef; luncheon meat (similar to emulsiontype sausage) made from beef, pork, or a mixture of beef and pork, or from other kinds of meat; pressed beef (made from large cured high-quality meat pieces; picnic ham (usually weighing 0.5 kg); canned cured ox tongues. Similar products made from pork are: minced pork, chopped pork, lunch pork, pressed pork, pressed ham and many others. Stewed beef, stewed mutton, stewed veal and stewed venison are frequently canned, as well as dishes in jelly such as veal in jelly. Liver sausage, liver paste and some other pastes and frankfurters in brine are also suited for canning.

## *Containers Suited for Sterilization of Meat Products and Hermetical Closure*

*Container.* The container protects the canned product from spoilage by recontamination with micro-organisms, therefore reliable containers and properly adjusted closing machines are essential to prevent access of spoilage organisms during the cooling operation and during the shelf-life of the product. Before filing, containers have to be rinsed or otherwise cleaned from dust or other impurities.

Tin-plate cans are most commonly used, such as the soldered side-seam three-piece can, consisting of can body, can end and lid. The cemented sideseam and welded side-seam cans constitute the majority of cans produced, and remain one of the most reliable and acceptable packages ever conceived. The benefits of using three-piece cans include no size limitations, a wide range of plate thicknesses and tempers for body and ends, abuse resistance, strong end profiles and long shelf-life. Two-piece cans are becoming popular for some applications. Can end and body is drawn in a single piece from flat sheet stock. Only shallow cans (height less than half the diametre) can be fabricated this way. Enamelled cans are more suitable for canned meat products because interior corrosion of cans can be avoided.

Increasingly more attention is being given to aluminium for manufacturing cans and other containers for canned meat products, where special opening features are desired. Important advantages of aluminium cans are that they are lead-free and do not rust. Most often two-piece aluminium cans are used in meat canning. Aluminium for can ends and bodies is, without exception, coated with enamel on both sides. It is widely used in flexible and semi-rigid containers as a protective packaging for a large number of meat products. As a result of extensive development, the use of flexible, laminated pouches and formed aluminium containers for shelf-stable sterilized products is a commercial reality.

The retortable pouch, a thin rectangular package used for heat-sterilized meat products, offers potential improvements in convenience and quality because of its shape and composition. During heat processing, its shape and size allows rapid heat transfer to destroy micro-organisms at the innermost part of the pouch without excessively overheating the product near the pouch wall. The contents are likewise more rapidly cooled at the end of sterilization. A suitable laminate used to make retortable pouches consists of three foils (polyester, aluminium and polypropylene film or resin). These containers are heat-sealed.

Glass containers. Glass is an inert container, although damage to the meat product may result from exposure to light. Glass jars are sealed with caps, twist-off lids and many other systems (crowns, side seal closures, rolled-on closure, screw caps or lug caps) and are used for frankfurters in brine and similar sausages and to some extent also for liver sausage mixtures or sausage mixtures with high-collagen content.

Retortable synthetic casings. Recently a new kind of casing has been used for sausage filling, especially if it is necessary to obtain a longer shelf-life for sausages, sterilized in a retort. This kind of container is sealed with clips made of aluminium.

Smaller size containers are more suitable for meat canning because heat penetration of the meat is by conduction, so if larger containers are used extremely severe heat treatments will be necessary for sterilization. Such severe heat treatment will result in a much lower quality of canned product which will be extremely overcooked. That is why larger pieces, such as hams, shoulders, etc. are pasteurized at temperatures of about 80°C.

*Hermetic closure*. Tin-plate cans and aluminium cans are closed by machines which form a double seam in three stages: pressing the cover (lid) to the can body, forming the first operation roll seam, and forming the second-operation roll seam.

Good double seams are absolutely essential to prevent losses due to spoilage. The food processor is responsible for the quality of seams produced in the plant and must provide adequately trained personnel to perform seam evaluations and make closing machine adjustments.

Making good seams requires careful closing machine maintenance, frequent, regularly scheduled seam evaluations and immediate correction of seam conditions which are outside established tolerances. Hermetic closure of cans under vacuum is highly recommended, particularly for countries with a hot climate.

### Autoclave (Retort)

Batch retorts are heated with steam or water under pressure. Saturated air-free steam is an excellent heating medium for processing canned meat products. Hot water under pressure has many of the advantages of saturated steam as a heating medium but it must be circulated among containers and continually reheated.

The canning operation must be performed only by fully trained personnel. Operation of batch retorts heated with steam is performed in the following steps:

- load the cans in trays or baskets and close and tighten the retort lid;
- open the drain, vent condensate valves and the high-and low-pressure steam supplies;
- remove the air by a flow of steam during venting;
- close drain and vent valves;
- bring retort to operating temperature and pressure;
- regulate steam supply by temperature control;
- time the process from when the retort reaches the specified temperature;
- turn off the steam at the end of the process;
- start cooling by flooding the retort with cold water for about ten minutes (smaller cans up to 500 grams capacity), or 20 minutes (larger cans up to 1 500 grams);
- continue cooling until the average temperature of the product is 40°C when the process is completed;
- open the lid of the retort, remove baskets with canned products and let them dry.

Venting is one of the most important operations in processing canned meat products in steam, because air remaining among cans in an inadequately vented retort forms an insulating layer and will reduce the efficiency of the process.

In general, the same principles apply to processing in water. Some containers such as glass jars and flexible synthetic pouches are usually retorted in hot water in vertical retorts under superimposed air pressure. Chlorination of cooling water is recommended (0.5 ppm free chlorine should be the lowest amount at the point of discharge).

### Frequent Faults Committed During Canning

*Insufficient sterilization* results in spoilage of canned product and is identified by can swell (hard, soft and springer swell) or by flat sour spoilage.

*Overprocessing* results in too soft a consistency and a burnt taste.

*Deformation of cans and semi-rigid containers, breakage of glass jars and bursting of casings* are the results of a careless, cooling operation, especially improper superimposed pressure.

*Leaking cans* are the result of improper closing machine adjustment and deformation of cans prior to closing or after being processed, especially near the double seam.

### Shelf-life

Fully preserved canned meat products keep up to four years, but in the tropics it is only about one year.

Canned pasteurized products should be clearly labelled as perishable product (keep under refrigeration or store at/below 5°C) and must not be confused with fully sterilized products which do not require refrigerated storage.

Most canned meat products are eaten after emptying the can, without reheating, stews and frankfurters being the exceptions.

## Meat and Meat Products

Meat is the most valuable livestock product. Meat is composed of protein and amino acids, minerals, fats and fatty acids, vitamins and other bioactive components, and small quantities of carbohydrates. From the nutritional point of view, meat's importance is derived from its high quality protein, containing all essential amino acids and it's highly bioavailable minerals and vitamins.

While meat consumption has been relatively static in the developed world, annual per capita consumption of meat has doubled since 1980 in developing countries. Growing population and incomes, along with changing food preferences are increasing the demand for livestock products.

World meat production is projected to double by 2050, most of which is expected in developing countries. The growing meat market provides a significant opportunity for livestock farmers and meat processors in these countries. Nevertheless, increasing livestock production and the safe processing and marketing of hygienic meat and meat products represents a big challenge.

The FAO programme in meat and meat products aims to assist the member countries in exploiting the opportunities for livestock development and poverty alleviation through the promotion of safe, efficient and sustainable production, processing and marketing of meat and meat products. The focus is on enhancing skills and capacity building in the smallholder sector through improving and upgrading small-scale meat production and processing techniques. FAO also assists with organising marketing and meat value chain improvement through a combination of headquarters-based and field activities and collaboration with selected national, regional and international partners.

The main emphasis is on adding value, improving safety, minimising waste and providing policy and technical advice and assistance. The approach is to develop and disseminate manufacturing guidelines and practices for improved productivity and safer, value added products and processing facilities. FAO is also involved, through the Codex Alimentarius, in the development meat and meat products standards and codes of practice.

## Meat Consumption

Meat can be part of a balanced diet contributing valuable nutrients that are beneficial to health. Meat and meat products contain important levels of protein, vitamins, minerals and micronutrients which are essential for growth and development. Further processing of meat offers the opportunity to add value, reduce prices, improve food safety and extend the shelf-life. This can result in increased household income and improved nutrition. While the per caput consumption of meat in some industrialized countries is high, per caput consumption below 10 kg in developing countries must be considered insufficient and often leads to under-nourishment and malnutrition. It is also estimated that more than 2 billion people in the world are deficient in key vitamins and minerals, particularly vitamin A, iodine, iron and zinc. Deficiencies occur when people have limited access to micronutrient-rich foods such as meat, fish, fruit and vegetables. Most people with micronutrient deficiencies live in low income countries and are typically deficient in more than one micronutrient. Highly nutritious foods such as meat are particularly required for HIV AIDS infected communities and also for women and children.

## World Meat Markets at a Glance (FAO World Food Outlook 2008)

To effectively combat such malnutrition and under-nourishment, 20 g of animal protein per person per day or 7.3 kg per year should be provided. This can be achieved by an annual consumption of 33 kg lean meat or 45 kg fish or 60 kg eggs or 230 kg milk, respectively. These sources are usually combined in the daily food intake, but in regions where not all of them are readily available, intake of the others needs to be increased. Although nutrients from animals may be of higher quality or more readily absorbed than vegetable sources, it is possible to have a healthy vegetarian diet.

The steadily growing world population and increasing incomes creates higher demand for meat, but at the same time leaves limited space for expansion in livestock production. Therefore the maximum utilization of existing food resources becomes even more important. Poultry meat is of growing importance to meet this demand.

The FAO programme in meat and meat products aims to assist the member countries in exploiting the opportunities for livestock development and poverty alleviation through the promotion of safe and efficient production, processing and marketing of meat and meat products.

## Sources of Meat

The most common sources of meat are domesticated animal species such as cattle, pigs and poultry and to a lesser extent buffaloes, sheep and goats. In some regions other animal species such as camels, yaks, horses, ostriches and game animals are also eaten as meat. To a limited extent, meat is also derived from exotic animals such as crocodiles, snakes and lizards.

For thousands of years, poultry supplied meat and eggs, cattle, sheep and goats provided meat and milk, and pigs provided a source of meat. These species are the main sources of animal protein for humans. The meat derived from cattle is known as beef, meat derived from pigs as pork and from chickens as poultry. Pork is the most widely eaten meat in the world accounting for over 36% of the world meat intake. It is followed by poultry and beef with about 33% and 24% respectively.

| | *1990* | *2000* | *2010* | *% change 1990-2010* |
|---|---|---|---|---|
| Bovine | 1447 | 1479 | 1623 | 12 |
| Porcine | 856 | 899 | 966 | 13 |
| Poultry | 11792 | 16078 | 21489 | 82 |
| Ovine | 1799 | 1811 | 2000 | 11 |

The utilization and subsequent consumption of different animal species vary as cultural preferences and religious beliefs are observed.

### Animal Quality

The quality of animals slaughtered has a big effect on the standard of meat produced. The most important factors are feeding, age, genetics and health status. Efforts have been made to improve meat production and quality by breeding, or to combine the key characteristics by means of crossbreeding. New breeds with improved meat quality and carcass yields and adapted disease resistance can make a significant contribution to improved meat availability for human nutrition. Typical cattle breeds originating from Asia are Brahman and Gyr. These breeds and their crosses are common in most tropical countries around the world. Angus, Charolais, Hereford, Limousin and Simmental are typical European breeds. Also the increasingly popular Japanese Wagyu is a cross of native cattle and European inputs.

Most authorities have considered that the origins of the domestic chicken (*Gallus domesticus*) were the Red Jungle Fowl of Asia. The present day chickens are larger and more productive than the ancestral Red Jungle Fowl with four basic types: 1) egg type, or Mediterranean; (2) game; (3) meat type; and (4) Bantam. Within these types, there are an estimated 1233 different breeds of chicken that are recognized. Most commercial broilers are hybrid chickens.

A huge number of native pig breeds exist around the world and are well adapted to local conditions. Typical commercial pig breeds are the various landraces (Danish, German, Dutch, Italian, etc.) and the Large Whites in Europe, often crossed with the Pietrain race from Belgium. In Asia, the Beijing Black and the Jinhua pig from China and the Mong Cai from Vietnam are common. Most commercial pigs are crosses or hybrids.

### Productions Systems

Farming is a complex, multicomponent, interactive process that is dependent on land, animal, human and water resources as well as capital investment. Throughout the world it is practiced in many different ways and environments and with differing degrees of intensity and biological efficiency. Animal production systems vary from mainly capital intensive in the western world to mainly capital extensive systems in the developing world. The FAO (Sere and Steinfeld, 1996) has defined 3 main livestock production systems; Industrial, Mixed and Grazing systems.

### Industrial Systems

Industrial livestock systems are those in which the animals are detached from the land base of feed supply and waste disposal. They depend on external supplies of feed, energy and other inputs. Industrial systems provide >50 percent of global pork and poultry meat production and 10 percent of beef and mutton production. Excess nitrogen and phosphorus nutrients due to large quantities of feed imports, can lead to problems with manure disposal and pose a potential pollution risk.

### Mixed Systems

In mixed farming systems, crops and livestock production are integrated on the same farm. Globally mixed farming systems produce the largest share of total meat (54 percent), and milk (90 percent). Mixed farming is the main system for smallholder farmers in many developing countries.

### Grazing Systems

Grazing systems are defined as livestock systems in which more than 90 percent of dry matter fed to animals comes from rangelands, pastures, annual forages and purchased feeds and less than 10 percent of the total value of production comes from non-livestock farming activities. In terms of total production, grazing systems supply only 9 percent of global meat production yet provide the sole source of income for 20 million pastoral families.

## Slaughtering Facilities

A slaughterhouse also called an "abattoir", is a facility where animals are killed and processed into meat products. In large facilities, slaughtering is carried out in fully mechanised lines. The workers are assigned to specific workstations and the carcasses move on a conveyor system from station to station until the slaughter process is completed.

In slaughter line operations, clean and unclean operations are physically separated and individually manned to avoid contamination of carcasses and edible by-products. The unclean operations include stunning, bleeding, and defeathering (poultry), dehairing (pigs) and dehiding (cattle and small ruminants). The clean operations include evisceration, carcass splitting and carcass dressing.

In many developing countries adequate slaughter facilities are not available. At rural or local level slaughtering is often either carried out under a tree or in deteriorated and outdated slaughter units without any waste treatment facilities. This often results in health hazards through contamination of the meat during slaughter operations and of the surrounding land and water through uncontrolled release of waste and effluents. In rural settings slaughtering can be done in basic small-scale facilities. Before building a slaughtering facility a number of factors must be considered including the species to be slaughtered, expected slaughter numbers, environmental impact, the availability of a competent workforce and the linkages to the meat markets.

To facilitate the efforts of governments and private investors to improve meat hygiene, the FAO/WHO Codex Alimentarius committee on meat hygiene has developed guidelines in the "Code of Hygienic Practices for Fresh Meat" on the design, facilities and equipment of slaughter establishments.

## Slaughtering Process

In many developing countries, lack of appropriate slaughtering facilities and unsatisfactory slaughtering techniques are causing unnecessary losses of meat as well as valuable by-products from animal carcasses. Therefore adequate facilities, clearly defined working procedures and well-trained skilled labour are essential to guarantee a safe supply of wholesome meat.

### *Slaughter Process Diagram*

In cattle slaughter commonly mechanical stunning (penetrative and non-penetrative) with a captive bolt pistol is used. Electrical stunning is increasingly an option. This is followed by bleeding in a hanging position. In smaller operations, the carcass is lowered onto a cradle, head and feet are removed and flaying (removal of skin) initiated. Then the carcass is raised again and flaying completed. In bigger operations, carcasses remain suspended on overhead rails and hide pulling devices are used to reduce contamination of the carcass surface. After flaying, evisceration takes place. White and red offal are separated, inspected and cleaned. Condemned material is separated and disposed of in a sanitary manner. The slaughter procedure for small ruminants is similar to that of cattle.

Poultry slaughter in larger operations, the birds are suspended by the leg on solid wire shackles attached to a conveyor system taking them to the various workstations. The birds are stunned prior to bleeding and then passed through either scalding water or a steam tunnel and through rotating devices with rubber scrubbers to remove feathers. After chilling the birds in iced water, evisceration takes place. In smaller operations, the birds are usually fixed in cones for bleeding, scalded in hot water and defeathered either manually or by use of a drum plucker. They are then eviscerated and chilled.

In pig slaughter electrical stunning is commonly used (carbon dioxide gas is an option). In smaller operations, the pig is hoisted for bleeding. After bleeding, the carcass is lowered into the scalding tank filled with water at around +60°C (140°F). After a few minutes, the carcass is removed and the hair is scraped off manually. After dehairing, the carcass is usually exposed to open gas flames to remove remaining hair and reduce surface contamination. The carcass is then transferred to the overhead rail where final cleaning takes place, followed by evisceration. The internal organs are inspected and cleaned for further use. The carcass is split and inspected and either chilled or directly dispatched. In larger operations pigs move on conveyor systems through scalding tunnels and dehairing machines with rotating rubber fixtures.

### Meat Cutting

In many developing countries meat is marketed fresh immediately after slaughter. With the development of societies, urbanization and growth in disposable income levels, the demand for meat and processed meat products is increasing. As a result, suitable cutting schemes have been developed, identifying and separating meat for direct sales and meat suitable for further processing. Originally these schemes differed based on local consumption patterns and cooking preferences. Over time and with growing local and regional trade, authorities have introduced suitable uniform schemes for domestic, regional and international markets.

## Processing Technologies

Meat processing technology consists of technologies and procedures in the manufacture of processed meat products. Processing of meat maximises the utilization of meat and slaughter by-products. Meat mixes containing lower grade meat trimmings and additional non-meat ingredients are a valuable source of animal protein in the diet. The animal tissues, muscle meat and fat are the main ingredients. Occasionally other tissues such as internal organs, skins and blood are used and complemented with ingredients of plant origin.

All processed meat products on the market have been physically and/or chemically treated. These treatments go beyond the simple cutting of meat into meat cuts or meat pieces with subsequent cooking as meat dishes. Modern meat processing involves a range of physical and chemical treatment methods. One single method can be applied but it is often a combination of various methods.

- Meat processing technologies include various individual steps such as:
- Cutting, grinding and chopping
- Mixing and tumbling
- Salting and curing for
- Application of spices and non-meat ingredients
- Stuffing of meat mixes into casings or other containers
- Drying and fermentation
- Hot or cold smoking
- Heat treatment (pasteurisation, sterilisation)

Basic meat processing can be carried out manually using simple tools and only limited equipment. Increasingly modern meat processing is more mechanized using specialized equipment and tools. The core processing equipment consists of a meat grinder, a bowl cutter, cooking vat, smokehouse and chiller. The most essential tools are brine pump, cutting table, butcher knives and bone saws. Such equipment and tools are available tailor-made for small-scale, medium-sized or large-scale operations.

FAO promotes the introduction of modern safer meat processing techniques. In the past this included initialising and conducting intensive hands-on training courses at national level or as part of regional projects. Recent activities include the regional technical coop programme (TCP) training in Africa (1997-1999) and Asia & the

Pacific (2003-2005). Several FAO publications are available with valuable in-depth information on meat processing.

### Product Groups

Around the world, a huge variety of semi-processed and processed meat products with different taste characteristics have evolved over time. In some regions hundreds of different meat products exist with individual names and taste characteristics. Many of these products use similar processing technologies despite different tastes and shapes. These products can be categorized as follows:

### Fresh Processed Meat Products

These products consist of raw meat and fatty tissue. Spices, common salt and sometimes also binders are added. In low-cost versions extenders or fillers are added for volume extension. Products are marketed raw, but fried or cooked prior to consumption to make the products palatable. If the fresh meat mixes are filled in casings, they are defined as sausages. If other portioning is customary, the products are known as patties or kebabs. Typical fresh products are merguez, longganisa, bratwurst, breakfast sausage, burger or suflaki.

### Cured Meat Cuts

Muscle pieces are used for these products which can be subdivided into cured-raw meats and cured-cooked meats. The curing process is similar for both groups. The meat is treated with small amounts of curing salt, either by simple dry-salting or by injecting and/or submerging in a curing salt solution. Cured-raw meats are not heat-treated and only subjected to curing, drying, fermentation and ripening. They are usually consumed raw. Typical products are Parma and Serrano ham. Cured-cooked meats always undergo heat treatment after a short curing process to achieve the desired palatability. Typical products are Virginia and sandwich ham.

### Raw-cooked Meat Products

In this product group, the muscle meat, fat and non-meat ingredients are first processed raw by grinding, chopping and mixing. The resulting viscous batter is portioned in sausages or loaves and then subjected to heat treatment, which causes protein coagulation, a firm-elastic texture, palatability and some degree of bacterial stability. Sausages are usually cooked or steamed and, when stuffed in permeable casings, also hot-smoked. The loaves are usually baked. Typical products are mortadella, hotdogs, frankfurters, viennas and meat loaf.

### *Precooked-cooked Meat Products*

Such products can contain mixes of lower-grade muscle trimmings, fatty tissues, head meat, animal skin, blood, liver and other edible parts. In general, two phases of heat treatment are involved in the manufacturing process. The first phase is the precooking of raw meat materials and the second phase the cooking of the final product mix. Precooked-cooked meat products utilize the greatest variety of meat, animal by-product and non-meat ingredients. Typical products are liver pates, blood sausages and corned beef.

### *Raw-fermented Sausages*

Raw-fermented sausages consist of coarse mixtures of lean meats and fatty tissues combined with curing salt, sugars, spices and other non-meat ingredients and are usually stuffed into casings. They get their characteristic flavour, texture and colour through fermentation in combination with moisture reduction. The final products are not subjected to heat treatment and distributed and consumed raw. Typical products are chorizo and salami type summer sausages.

### *Dried Meat Products*

These products are the result of simple drying of lean meat. Their processing is based on the experience that meat will not easily spoil when a substantial part of the natural tissue fluid is evaporated. Mostly pieces of lean meat are cut to a specific uniform shape that permits the gradual and equal drying of whole batches. Dried meat has a significantly longer shelf-life than fresh meat. The nutritional value of the protein content remains unchanged. Typical products are jerkey, biltong, charque, pastirma.

### *Basic Hygiene Practice*

***Control of Hygiene and Safety :*** Ensuring the safety of meat calls for control throughout the food chain from the farm of origin, inspection before and after slaughter, to the handling and storage of meat and the products until the time it is consumed.

The Codex Alimentarius commission has elaborated (besides meat inspection Codes) the Recommended International Code of Hygiene Practice for Fresh Meat and the Recommended International Code of Hygienic Practice for Poultry Processing which describe the minimum hygiene requirements for safe meat and poultry production.

- The application of these Codes can be an important step towards the targets:

- That the food will not cause infection or intoxication when properly prepared;
- Does not contain residues (of pesticides, veterinary drugs and heavy metals) in excess of established limits;
- Is free from disease;
- Free from obvious contamination;
- Free from defects generally recognised as objectionable;
- Has been produced under adequate hygienic control;
- Fulfils the expectation of the consumer in regard to composition.

The Codex Alimentarius commission guidelines include advice on the construction of abattoirs and the facilities required; control of pests, quality of water for cleaning and disinfection; rules of meat inspection and hygienic practices (including supervision by a veterinary inspector).The responsibility for the production of safe and wholesome meat is normally shared by the industry and the controlling authority. This requires a controlling authority which is adequately resourced and has the legal power to enforce the requirements and which should be independent of the management of the establishment where the meat is produced.

## Good Manufacturing Practice

A fast growing population in developing regions and the tendency to live in densely populated areas and mega cities increases the risk of rapid disease transmission and large-scale outbreak of food-borne diseases due to the increasing length and complexity of supply chains. One source of such cross contamination can be the slaughterhouse and it is therefore mandatory to observe a stringent regime of control. Meat inspection consists of ante-mortem inspection of slaughter animals and post-mortem inspection of slaughtered animals. These are procedures conducted for the purpose of judgement of safety and suitability and disposition. They play a vital role as main control measures to assure meat safety and of course to identify and monitor animal diseases.

The inspection is usually carried out by designated veterinarians or meat inspectors at the abattoir. The meat inspection procedure is carried out in two phases, ante and post mortem inspection. During ante mortem inspection the slaughter animals are screened to assess general behaviour, nutritional status, cleanliness, signs of diseases and abnormalities in posture, structure, colour, discharges and odour. Suspect dirty, diseased, and injured animals are immediately separated from the healthy animals to avoid cross contamination and slaughtered separately. Routine post-mortem examination is routinely carried out to ensure that the carcasses and its organs are disease free. Results from ante and post mortem inspection are then combined to make a final judgement on suitability of the meat or organs for human consumption. FAO has conducted training programmes for meat inspectors and prepared concise guidelines on meat inspection of bovine, small ruminants, pigs, game, poultry and rabbits.

### Haccp

Hazard Analysis and Critical Control Point (HACCP), is used to descibe an internationally recognised way of managing food safety and protecting consumers.

- The HACCP system is a preventative approach to food safety based on the following 7 principles:
- Identify any hazards that must be prevented eliminated or reduced
- Identify the critical control points (CCPs) at the steps at which control is essential
- Establish critical limits at CCPs
- Establish procedures to monitor the CCPs
- Establish corrective actions to be taken if a CCP is not under control
- Establish procedures to verify whether the above procedures are working effectively
- Establish documents and records to demonstrate the effective application of the above measures

It is generally accepted that HACCP is the most effective means of ensuring the safety of food, thereby protecting consumer health.

### Traceability

The Codex Alimentarius Commission defines Traceability as "the ability to follow the movement of a food through specified stage(s) of production, processing and distribution". Livestock identification and meat traceability systems were introduced in countries as a reaction to the BSE problem. A good traceability system provides accurate information on the origin, sex, age, breed, movements and records veterinarian treatments an animal received. Traceability systems are important tools to prevent the spread of animal diseases and to enhance biosecurity in general.

Animal identification and livestock/meat traceability are not themselves food safety, animal disease prevention or quality assurance programmes, but they facilitate public health, veterinary public health and animal health interventions. FAO through its contribution to the Codex Alimentarius commission is involved in inclusion of traceability within international standards.

Common animal identification options include:

- Paint/spray markings
- Hot/Cold branding
- Tattoo
- Collar tag (neck band)
- Plastic/metal eartags
- Microchip (RFID-Radio frequency identification device)

Within the abattoir, meat traceability requires the clear identification of the live animals through to meat products. Possible meat identification options include

- Paper/plastic Tags
- Ink stamping
- Bar code labels
- Microchip (RFID)

Costs of the identification and thereby traceability systems vary hugely, depending on the options applied and the level of detail required. Functioning traceability schemes are often a precondition for meat producing countries to enter export markets. Hence traceability is not only a health issue but has also become a marketing tool. The goal is to achieve transparency in the meat chain "from farm to fork".

### Meat Quality

Meat quality is normally defined by the compositional quality (lean to fat ratio) and the palatability factors such as visual appearance, smell, firmness, juiciness, tenderness, and flavour. The nutritional quality of meat is objective yet "eating" quality, as perceived by the consumer, is highly subjective.

### Visual Identification

The visual identification of quality meat is based on colour, marbling and waterholding capacity. Marbling is small streaks of fat that are found within the muscle and can be seen in the meat cut. Marbling has a beneficial effect on juiciness and flavour of meat. Meat should have a normal colour that is uniform throughout the entire cut. Beef, lamb, and pork should also have marbling throughout the meat.

### *Smell*

Another quality factor is smell. The product should have a normal smell. This will be different for each of the species (i.e. beef, pork, chicken), but should vary only slightly within the species. Any rancid or strange smelling meat should be avoided.

### *Firmness*

Meat should appear firm rather than soft. When handling the retail package, it should be firm, but not tough. It should give under pressure, but not actually be soft.

### *Juiciness*

Juiciness depends on the amount of water retained in a cooked meat product. Juiciness increases flavour, helps soften meat - making it easier to chew, and stimulates saliva production in the mouth. Water retention and lipid content determine juiciness. Marbling and fat around edges helps hold in water. Water losses are from evaporation and drip losses. Meat aging can increase water retention and therefore increases juiciness.

### *Tenderness*

Has been linked to several factors, such as the animal's age, sex or the muscle location. One important way to tenderize meat is by aging. Carcasses are aged by holding them at refrigeration temperatures for extended periods of time after slaughter and initial chilling.

### *Flavour*

Flavour and aroma are intertwined to create the sensation the consumer has during eating. These perceptions rely on the smell through the nose and on the sensations of salty, sweet, sour and bitter on the tongue. Meat flavour is affected by type of species, diet, cooking method and method of preservation (e.g. smoked or cured).

### *Public Health Issues*

Although meat is generally perceived as a high value product which is a good source of iron and vitamins, globally there have been a number public health issues which have had adverse effects on the production, trade and consumption of meat. The risk of animal zoonoses is perhaps the biggest cause of public concern with meat. Zoonoses are defined by the WHO as "Any disease or infection which is naturally transmissible from vertebrate animals to man". There are over 200 zoonoses known to man of which tuberculosis alone causes more deaths worldwide than any other infectious disease, with an estimated 1.6 million deaths in 2005.

Meat inspection is the principal method for ensuring the safety of meat. The purpose of meat inspection is to provide safe and wholesome meat for human consumption. The responsibility for achieving this objective lies primarily with the relevant public health authorities who are normally represented by veterinarians or meat inspectors at the abattoir. Common weaknesses in meat inspection include a lack of properly resourced, independent and a competent body of inspectors (veterinarians, meat inspectors) who are not fully legally and practically empowered to implement their objective judgement on the fitness of meat and animal parts for human consumption.

FAO has also produced guidelines and standards on meat inspection procedures and techniques for developing countries. Capacity building in updated meat inspection has also been a key feature of FAO support to member countries. FAO and the World Health Organization (WHO), in addition to the World Organisation for Animal Health (OIE), is responding to the global threats of animal zoonoses through surveillance and research networks for early detection of animal diseases and better scientific cooperation between countries. FAO has also established an emergency crisis management centre (CMC) that supports countries in responding to animal disease outbreaks.

# Bibliography


Anderson, Warren R.: *Mastering the craft of smoking food,* NJ, Burford Books, Short Hills, 2006.

Andrea, A. Louise: *Home canning, drying and preserving,* Page & Company, Garden City, NY, Doubleday, 1918.

Bell, Mary: *Food drying with an attitude: a fun and fabulous guide to creating snacks, meals, and crafts,* Skyhorse Pub., New York, 2008.

Berolzheimer, Ruth: *Victory canning: preserving, drying, smoking and pickling of fresh foods for future use,* Consolidated Book Publishers Inc., Chicago, 1942.

Chadwick, Janet: *The beginner's guide to preserving food at home,* Storey Pub., North Adams, MA, 2009.

Collins, James H.: *The story of canned foods,* E. P. Dutton & Co., New York, 1924.

Dubbs, Chris, and Dave Heberle: *Smoking food: a beginner's guide,* Skyhorse Pub., New York, 2008.

Durham, T. R.: *The smoked seafood cookbook: easy, innovative recipes from America's best fish smokery,* University of Michigan Press, Ann Arbor, 2008.

Emery, Carla, and Lorene Edwards Forkner: *Canning & preserving your own harvest: an encyclopedia of country living guide,* Sasquatch Books, Seattle, WA, 2009.

Geary, Kelly, and Jessie Knadler: *Tart & sweet: 101 canning recipes for the modern kitchen,* Rodale, Emmaus, PA, 2010.

Greensmith, Maurice: *Practical dehydration,* CRC Press, Boca Raton, 1998.

Hill, Janet McKenzie, *Canning, preserving and jelly making,* Brown and Company, Boston, Little, 1915.

Hobson, Phyllis: *Making & using dried foods,* Storey Communications, Pownal, VT, 1994.

Hui,Y. H.: *Handbook of animal-based fermented food and beverage technology,* CRC Press, Boca Raton, 2012.

Keith H. Steinkraus: *Industrialization of indigenous fermented foods,* Marcel Dekker, New York, 2004.

Lewin, Alex.: *Real food fermentation: preserving whole fresh food with live cultures in your home kitchen*, Quarry Books, Beverly, MA, 2012.

Livingston, A. D.: *Cold-smoking & salt-curing meat, fish & game,* Lyons Press, Guilford, CT, 2010.

Marianski, Stanley and others: *Meat smoking and smokehouse design,* Bookmagic, Seminole, FL, 2009.

Marianski, Stanley, and Adam Marianski: *Home production of quality meats and sausages*, Bookmagic, Seminole, FL, 2010.

Marianski, Stanley, and Adam Marianski: *The art of making fermented sausages*, Bookmagic, Seminole, FL, 2009.

Megyesi, Jennifer Lynn: *The joy of keeping a root cellar: canning, freezing, drying, smoking, and preserving the harvest,* Skyhorse Pub., New York, 2010.

Neil, Marion Harris: *Canning, preserving and pickling,* D. McKay, Philadelphia, 1914.

Oster, Kenneth V.: *The complete guide to preserving meat, fish, and game: step-by-step instructions to freezing, canning, and smoking,* Atlantic Pub. Group., Ocala, FL, 2011.

Popeil, Ron, and John May: *Dehydrated & delicious: the complete book on dehydrating meats, fruits, vegetables, herbs, flowers, yogurt, and more,* Dehydrator Products, Carlsbad, CA, 1992.

Rombauer, Irma von Starkloff, and others: *Joy of cooking: all about canning & preserving*, Scribner, New York, 2002.

Rorer, Sarah T.: *Canning and preserving,* Arnold and Co., Philadelphia, 1887.

Ruhlman, Michael, and Brian Polcyn: *Charcuterie: the craft of salting, smoking, and curing,* W. W. Norton, New York, 2005.

Shafiur Rahman, M.: *Handbook of food preservation*, Boca Raton, CRC Press, 2007.

Shephard, Sue: *Pickled, potted, and canned: how the art and science of food preserving changed the world,* Simon & Schuster, New York, 2006.

Tamang, Jyoti Prakash: *Fermented foods and beverages of the world,* CRC Press, Boca Raton, 2010.

Thorne, Stuart: *The history of food preservation,* Barnes & Noble Books, Totowa, NJ, 1986.

White, Joanna: *The dehydrator cookbook,* Bristol Pub. Enterprises, San Leandro, CA, 1998.

Wilson, C. Anne: *Waste not, want not: food preservation from early times to the present day,* Edinburgh University Press, Edinburgh, 1991.

# Index


## A

Abnormalities, 63, 145, 312.
Additives, 26, 30, 47, 51, 55, 96, 128, 132, 136, 149, 257, 269, 273, 275, 277, 279, 287, 295, 296, 297.
Air Cell, 144, 162, 169, 173, 180, 181, 182.
Autoclave, 36, 37, 57, 105, 107, 266, 297.

## B

Beef Carcass, 7, 14, 23, 201, 240.
Biltong, 28, 32, 33, 34, 59, 78, 114, 316.
Burgers, 102, 111, 190, 258, 281.

## C

Candling, 144, 159, 160, 164, 172, 173, 174, 175, 176.
Casings, 53, 99, 105, 249, 270, 272, 277, 278, 279, 280, 283, 284, 286, 287, 290, 291, 294, 295, 296, 297, 300, 302, 309, 310, 311.
Cattle, 3, 5, 67, 68, 71, 74, 75, 80, 87, 88, 113, 123, 212, 213, 214, 222, 228, 230, 264, 277, 304, 305, 307, 308.
Common Salt, 18, 252, 260, 278, 283, 310.

## D

Dish, 26, 40, 106, 140, 149, 177, 178, 179, 184, 189, 190, 191, 192, 193, 195, 196, 197, 198, 229, 262, 295, 299, 309.
Dried Meat, 7, 8, 12, 13, 18, 24, 25, 26, 27, 28, 29, 30, 32, 40, 47, 50, 59, 61, 78, 104.
Dry Sausages, 26, 27, 40, 47, 58, 59, 61, 63, 99, 251, 253, 282, 284, 285, 286, 287, 288.

## E

Egg, 18, 66, 73, 100, 110, 121, 126, 128, 129, 130, 131, 132, 134, 136, 138, 140, 176, 177, 178, 179, 180, 200, 222, 269, 306, 307.
Egg-laying, 127, 128, 133, 138, 151.
Evaporation, 12, 13, 24, 46, 48, 58, 61, 155, 169, 224, 315.
Evisceration, 2, 3, 6, 77, 94, 136, 215, 278, 307, 308.

## F

Fatty Tissues, 14, 100, 248, 250, 251, 270, 271, 275, 276, 287, 288, 290, 296, 311.
Food Poisoning, 38, 43, 65, 80, 97, 125, 141, 206, 207, 208.
Freezing, 52, 56, 77, 93, 94, 95, 110, 200, 225, 281.

**G**

Grazing Systems, 306.
Growth Hormones, 136.

**H**

HACCP, 82, 83, 313.
Heat Treatment, 7, 30, 36, 38, 61, 96, 106, 107, 257, 258, 291, 293, 297, 301, 311.

**I**

Industrial Systems, 69, 306.
IIonising Irradiation, 91, 108.

**K**

Kilishi, 31, 32, 78, 104.
Kofta, 190, 191, 193, 198.

**L**

Labelling, 148, 154, 155, 175.
Luncheon Meat, 97, 98, 100, 106, 188, 299.

**M**

Meat Comminution, 269, 272.
Meat Preservation, 1, 2, 4, 7, 30, 52.
Meatball, 189, 190, 191, 192, 193, 194, 195, 196, 197, 198.

**N**

Natural Casings, 99, 272, 278, 290.
Nutritional Value, 62, 69, 89, 100, 138, 144, 162, 163, 311.

**O**

Organoleptic, 12, 26, 29, 40, 43, 44, 47, 51, 55, 256, 288.
Organoleptic Evaluation, 51, 52.

**P**

Pigs, 3, 5, 6, 7, 63, 67, 79, 85, 88, 99, 124, 210, 211, 212, 215, 224, 248, 249, 295, 296, 297, 307, 308, 313.
Poultry Farming, 126, 127, 128, 132.

**Q**

Quartering, 231.
Qwanta, 31, 78, 104.

**R**

Refrigeration, 7, 29, 33, 36, 45, 46, 47, 49, 59, 71, 72, 73, 184, 200, 264, 302, 315.
Riboflavin,85, 87, 92, 95, 99, 102, 108, 117, 143.

**S**

Sheep, 5, 7, 14, 67, 71, 88, 103, 201, 211, 212, 213, 214, 218, 224, 277, 278, 304.
Small Ruminants, 6, 307, 308, 313.
Soy Proteins, 273, 274, 275, 280.

**T**

Thermal Treatment, 36, 37, 38, 40, 41, 43, 44, 45, 267.
Traceability, 313, 314.

**V**

Vertical Method, 216, 218, 219.
Vitamins, 67, 84, 87, 89, 90, 91, 92, 94, 95, 102, 104, 107, 142, 143, 303, 304, 315.

**W**

Water Activity, 27, 34, 40, 44, 49, 58, 59, 60, 61, 102, 103, 104, 259, 272.
World Chicken Population, 137.

**Y**

Yeast Extract, 277.
Yolk, 100, 128, 138, 139, 140, 165, 167, 174, 185, 186, 198.


❑❑❑